INORGANIC MATERIALS CHEMISTRY:
GENERAL CONCEPTS AND RESEARCH TOPICS

INORGANIC MATERIALS CHEMISTRY:
GENERAL CONCEPTS AND RESEARCH TOPICS

BY
Dr. EKAMBARAM SAMBANDAN

S_ekambaram69@hotmail.com

iUniverse, Inc.
New York   Bloomington

Inorganic Materials Chemistry:
General Concepts and Research Topics

Copyright © 2008 by Dr. EKAMBARAM SAMBANDAN

All rights reserved. No part of this book may be used or reproduced by any means, graphic, electronic, or mechanical, including photocopying, recording, taping or by any information storage retrieval system without the written permission of the publisher except in the case of brief quotations embodied in critical articles and reviews.

The views expressed in this work are solely those of the author and do not necessarily reflect the views of the publisher, and the publisher hereby disclaims any responsibility for them.

iUniverse books may be ordered through booksellers or by contacting:

iUniverse
1663 Liberty Drive
Bloomington, IN 47403
www.iuniverse.com
1-800-Authors (1-800-288-4677)

Because of the dynamic nature of the Internet, any Web addresses or links contained in this book may have changed since publication and may no longer be valid. The views expressed in this work are solely those of the author and do not necessarily reflect the views of the publisher, and the publisher hereby disclaims any responsibility for them.

ISBN: 978-0-595-53173-8 (pbk)
ISBN: 978-0-595-63233-6 (ebk)

Printed in the United States of America

DEDICATED TO MY PARENTS

Preface:

I would like to introduce my book to readers with emphasis on basic and research needed in inorganic materials chemistry. There are two main topics included in the book and these are inorganic and materials chemistry. There are several sub-topics on the two main topics. All the topics consist of basic chemistry, and research topics needed are included as and when required. It is essential to understand what you read and think about it to generate new ideas. This was how I have got research fields required in the respective topics. Power of understanding the subject, thinking in successful direction and interest to work are required. I hope all the readers would certainly get into inorganic and materials chemistry field as a life career after reading this book.

My sincere gratitude goes to my wife, Kalyani and my daughter, Nandhini for their encouragement in writing this book. Also, I want to acknowledge my professors, relatives, friends and so on for their support on my career.

I can be reached at s_ekambaram69@hotmail.com for your comments and for the improvement of this book.

About the Author:

Dr. Ekambaram Sambandan has been devoted his time sincerely to work in the area of inorganic materials chemistry since he joined as a graduate student at Inorganic and Physical Chemistry department, Indian Institute of Science, India in August 1991. He has several years of Postdoctoral and industrial experiences in Japan, USA and South Africa. He has more than 20 publications and about 380 citations to his credit.

Drs. Keiters' work and encouragement

Contents

Inorganic Chemistry
Structure Of Atoms And Molecules 2
Sulfur Chemistry 21
Boranes Chemistry 28
Keto-Enol Tautomerism .30
Coordination Complexes .34

Materials Chemistry

Synthesis Of Inorganic Materials
Conventional Solid State Method: 40
Solid State Metathesis Method (Ssm): 45
Igneous Combustion Method: 47
Hydrothermal Method: 51
Ionothermal Method: 61
Ion-Exchange And Intercalation Methods: 62
Reverse Micelle Or Micro Emulsion Method 65
Precursor Method: 67
Co-Precipitation Method: .70
Microwave SyntHesis And Sintering Of Inorganic Materials: 72
Sol-Gel Method: 74
Electrochemical Methods: 79

Characterization Techniques
Powder Xrd 84
Simultaneous Tg-Dta: 88
Scanning Electron Microscopy: 89
Transmission Electron Microscopy: 89
Optical Spectroscopy: 89

Structure Of Inorganic Materials
Structure Of Nacl: 93
Structure Of Cscl: 93

Structure Of Zinc Blende:. 94
Structure Of Wurtzite: 95
Structure Of Fluorite: 96
Structue Of Rutile:. 97
Structure Of Perovskite: 97
Structure Of Spinel: 101
Special Structures Of Inorganic Oxides:. 101
   Important Topics
Photocatalysts For Water Splitting 108
Diluted Magnetic Semiconductors 123
Inorganic Phosphors For Solid State Lighting 126
Phosphors For Plasma Display Panels 128
Phosphor Materials In Fluorescent Lighting 131
Carbonitride With Sp^3 Carbon. 140
Open Framework Materials 143
Three Way Catalysts 145
Fuel Cell Materials 148
Thermoelectric Materials 152
Bioceramics 155
The Hydrogen Economy: The Forever Fuel, H_2 157
   Research Needed In Photosplitting Of Water
Igneous CombUstion Synthesis Of Oxide Semiconductors And Their
   Use For The Photosplitting Of Water At Room Temperature . . 161
Photosplitting Of Water Using Nanocomposite Semiconductors With
   Open Framework Structure 168
Design, Synthesis, And Photocatalysis Of Semiconductors With Open
   Framework Structure. 173
Igneous Combustion Synthesis, Characterization And Visible Light
   Driven Water Splitting Using Transition Metal Ion Incorporated
   Semiconductors 175
Mvo_4, M = Al, Bi, Cr, Fe, Y, Eu With Open Framework Structure. 176
   RelevAnt References: 177
      Appendix-1: 183

INORGANIC CHEMISTRY

CONCEPTS OF QUANTUM CHEMISTRY AND ITS APPLICATION TO STRUCTURE OF ATOM AND MOLECULE

Need:

Classical mechanics is perfect to explain macroscopic properties. However, it fails to demonstrate microscopic properties. Therefore, quantum chemistry is needed and it shows perfect fit for explaining microscopic properties such as structure of atom. In this section, we will see basic principle of quantum chemistry and its application to atomic and molecular structures.

Electromagnetic Radiation:

Electromagnetic radiation is the energy that travels in space. For example, ultraviolet and visible light come from the sun to the earth. It is characterized by three primary parameters. One is wavelength (λ), which is described by the distance between two consecutive peaks or trough in a wave. Another is frequency, υ defined by the number of waves per second that pass a given point in space. The third one is speed. Wavelength and frequency are related inversely. Thus, if wavelength is more, frequency is less and vice versa.

All the radiations are electromagnetic radiations and these radiations consist of electrical and magnetic vectors. The electrical vector is perpendicular to magnetic vectors. And these fields travel in space. The electromagnetic spectrum is given below.

		Wavelenth, nm			
10^{-11}	10^{-7}	10^{-5}	10^{-3}	10^{1}	10^{2}
A	B	C	D	E	F

Where
A = Gamma rays, B = X-rays, C = UV and Vis, D = IR, E = Micro wave and F = Radio frequency

Gamma rays have the lowest wavelength (~10^{-11} nm) whereas radio waves have the highest wavelength (~10^2 nm).

Quantization of radiations:

Max Planck first observed quantization of thermal radiation. This phenomenon was found in black body radiation. Black body is also considered having discrete energy levels. The energy of radiations is described by Planck equation.

$$\Delta E = nh\upsilon$$

Where h is Planck constant = 6.626×10^{-34} J

υ is frequency

n is an integer (1, 2, 3, ------)

The quantization of thermal energy, proposed by Max Planck, has been extended to all types of radiation by Einstein using his famous photoelectric experiment. Later De Broglie derived a four steps equation by comparing Planck's and Einstein's equations to show the dual property of electromagnetic radiation. Noble prize was awarded to De Broglie for his equation and the derivation and explanation of De Broglie equation are given below.

Max Planck equation is

$$E = h\upsilon = hc/\lambda$$

Einstein's equation is

$$E = mc^2$$

These two equations describe total energy and therefore, these two equations are compared.

$$hc/\lambda = mc^2$$

or

$$hc/\lambda = p.c$$

or

$$\lambda p = h$$

This equation is called De Broglie equation and according to this equation, if λ is more, wave nature is observed. If p is more, particle nature is observed. This concept is well understood by the following example. When tiny particle, for example electron, moves very fast, it behaves as wave.

The atomic spectrum of hydrogen:

Quantization of oscillators in black body is further extended to atomic energy levels. All the atoms have their own discrete energy levels. Hydrogen atom spectrum is a line spectrum and this observation of atomic spectrum can be explained by considering quantization of atomic energy levels.

High voltage   Excited by high voltage   Emission spectrum
H_2 → H → H* → H + Line spectrum
Low
Voltage

The line spectrum of hydrogen atom is obtained by high voltage discharge of low-pressure hydrogen molecule gas. The electric discharge breaks the hydrogen-hydrogen bond into hydrogen atom followed by excitation of hydrogen atom. When the hydrogen atom returns to ground state, it emits characteristic line spectrum. The electronic energy levels of hydrogen atom are represented in Fig. 1. Since hydrogen atom is an isolated sphere it can have only electronic energy levels. It shows that energy levels of hydrogen atom are compressed with increasing energy.

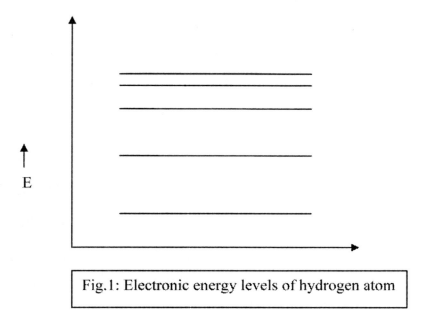

Fig.1: Electronic energy levels of hydrogen atom

Bohr model for hydrogen atom:

Bohr assumed fixed orbit motion of electrons around nucleus and each orbit has a discrete energy levels. This structure of hydrogen atom clearly explains the line spectrum of hydrogen atom. Thus, Bohr model of hydrogen atom is described below in Fig.2.

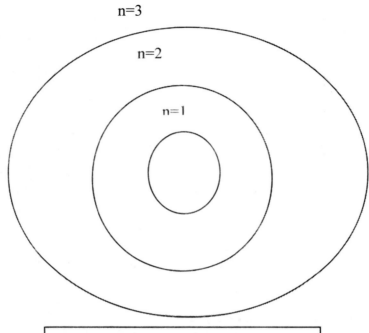

Fig.2: Bohr model of hydrogen atom

There is a strong attractive force between positive nucleus and negative electron. But, electrons do not fall in nucleus. This is due to centrifugal force experienced by revolving electrons, which keep electrons away from nucleus. Thus, at a fixed distance away from nucleus electrons revolve around nucleus (please refer to the above diagram to understand fixed orbits).

Quantum mechanical approach for the atom:

According to De Broglie, when tiny particle like electron moves faster, it behaves as wave. Therefore, it is assumed that electron around nucleus makes wave behavior. It is also important to note that the wave nature of electron is a standing wave. Therefore, electron can have a fixed energy level because of standing waves. The energy of electron is described by Schrodinger's equation.

$$H\Psi = E\Psi$$

Where H = Operator, Ψ = Wave function and E = Energy

For each wave function it will have different total energy. The wave function for the lowest hydrogen atom electron is 1s orbital.

Heisenberg uncertainity principle:

The product of uncertainity in position and in linear momentum is greater than or equal to $h/4\Pi$. Thus, the mathematical expression is

$$\Delta x . \Delta p > h/4\Pi$$

It means that if error (uncertainity) in determining the position of electron is less, then error in determining of p is more and vice versa.

Using a particle wave function, it is possible to determine the probability of finding electron. This is achieved by squaring the wave function. For example, the probability of finding 1s electron for the hydrogen atom is described using the following graph. (Fig.3)

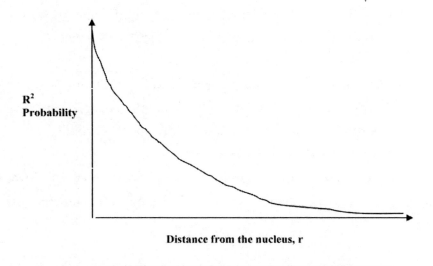

Fig.3: Probability of finding 1s electron from it nucleus in hydrogen atom

This plot indicates that probability of finding 1s electron decreases with distance from nucleus. It will be clear if we consider the radial probability distribution.

Radial Probability Distribution:

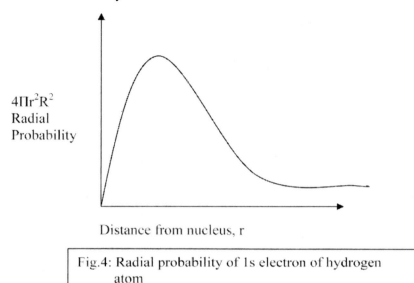

Fig.4: Radial probability of 1s electron of hydrogen atom

For the 1s orbital of hydrogen atom, the maximum radial probability occurs at a distance 0.529 Å from the nucleus.

Quantum Numbers:

Quantum numbers describe various properties of orbital of atoms. And hence, quantum numbers tell energy, orbital and spin of electrons around the nucleus. They are described one by one.

Principal Quantum Numbers (n):

Principal quantum numbers have integer, n = 1, 2, 3, ---. Each number tells about energy of orbital. With increasing 'n' radius of orbital increases. Therefore, electrons will stay away from the nucleus with increasing 'n'. Electrons have less energy for the higher principal quantum number.

Angular Momentum Quantum Numbers (l):

It has integral value from 0 to n-1. Thus, for n = 1, it has value 0, which is 1s orbital. The value of l tells us the shape of atomic orbitals.

Magnetic Quantum Numbers (m_l):

This has integral values from –l to l including zero. This quantum number tells us the orientation of the orbitals.

The following table 1 summarizes the quantum numbers for the first four levels of orbitals in the hydrogen atom.

Table 1: Quantum numbers of hydrogen atom

n	l	Orbital designation	m_l	Number of orbitals
1	0	1s	0	1
2	0	2s	0	1
	1	2p	-1, 0, 1	3
3	0	3s	0	1
	1	3p	-1, 0, 1	3
	2	3d	-2, -1, 0, 1, 2	5
4	0	4s	0	1
	1	4p	-1, 0, 1	3
	2	4d	-2, -1, 0, 1, 2	5
	3	4f	-3, -2, -1, 0, 1, 2, 3	7

Electron Spin Quantum Numbers (m_s):

It can have only one of two values, + ½ and – ½. Thus, the electron can spin in one of two opposite directions.

Orbital Shapes:

's' orbital:

The 's' orbital is spherical in shape. The centre of sphere is nucleus and at a particular distance from nucleus, electrons are found. Therefore, with increasing principal quantum number, n orbitals are progressively larger shapes. For example, 1s and 2s orbitals are shown in the following diagrams (Figs. 5 & 6).

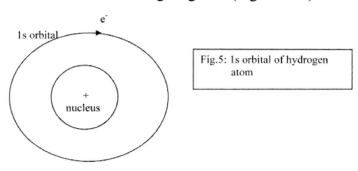

Fig.5: 1s orbital of hydrogen atom

n = 1 and l = 0.

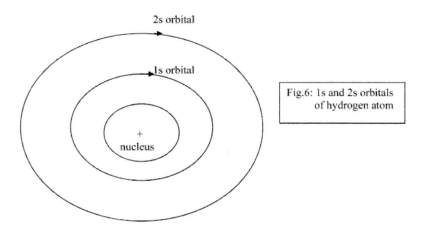

Fig.6: 1s and 2s orbitals of hydrogen atom

n = 2 and l = 0

The probability of finding electrons between 1s and 2s is zero and these regions are called node. In all the 's' orbitals, magnetic quantum number is zero.

'p' orbitals:

p orbitals consist of three orbitals of two lobed arrangement and these orbitals are p_x, p_y and p_z. p orbitals have three magnetic quantum numbers, +1, 0, -1. The orientation of p orbitals is shown below in Fig.7.

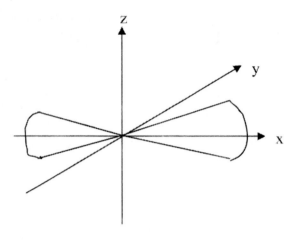

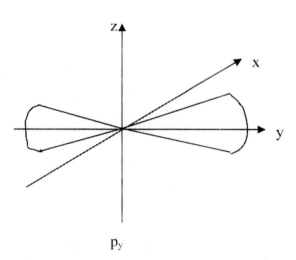

p_y

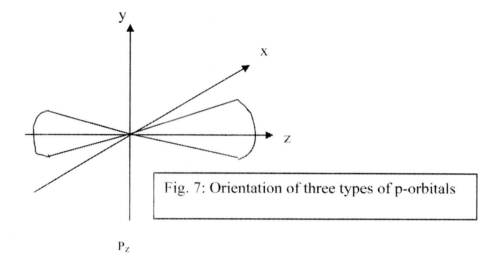

Fig. 7: Orientation of three types of p-orbitals

P_z

d-orbitals:

The d-orbitals consist of five orbitals with each one having four-lobed arrangement. The magnetic moments of these orbitals are -2, -1, 0, 1, 2.

Pauli Exclusion principle:

In a given atom no two electrons can have the same set of four quantum numbers (n, l, m_l and m_s).

Aufbau principle:

As protons are added one by one to the nucleus to build up the elements, electrons are similarly added to the hydrogenlike orbitals.

Hund's Rule:

The lowest energy configuration for an atom is the one having the maximum number of unpaired electrons allowed by Pauli principle in a particular set of degenerate orbitals.

The above mentioned principles are very much useful to derive electronic configuration of various ground state elements.

Electron arrangements in the atoms:

The electron arrangements in atoms depend upon the atomic number. One can do the electron arrangements in atoms by considering the following diagram (Fig.8).

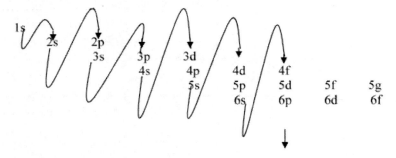

Fig.8: Electronic energy levels of atoms with increasing atomic numbers

s orbital can have maximum electrons of two.
p orbital can have maximum electrons of six.
d orbital can have maximum electrons of ten.
f orbital can have maximum electrons of fourteen.
e.g.:

Potassium, atomic number = 19.
Therefore the electronic configuration of potassium is
$1s^2$ $2s^2$ $2p^6$ $3s^2$ $3p^6$ $4s^1$

Valence electrons:

Valence electrons are the electrons in the outermost principal quantum level of an atom. The inner electrons are known as core electrons. The elements in the same group have the same valence electron configuration.

Atomic properties:

Atomic properties such as ionization energy, electron affinity and atomic size are described below with respect to atoms.

(i) Ionization energy:

The ionization energy is the energy required to remove the outermost electron of a gaseous atom. The removal of first electron is called first ionization energy and similarly, second, third and so on removal of electrons are called second, third and so on ionization energy. It is also noticed that second ionization energy is greater than that of first ionization energy and this is because of the following reason.. The first removal of electron of atom is from neutral atom. Whereas, the second removal of electron is from +ve ion. It is not easy to remove electron from +ve ion.
e.g.

$Al(g) \longrightarrow Al^+(g) + e^-$ $I_1 = 580$ KJ/mol

$Al^+(g) \longrightarrow Al^{2+}(g) + e^-$ $I_2 = 1815$ KJ/mol

$Al^{2+}(g) \longrightarrow Al^{3+}(g) + e^-$ $I_3 = 2740$ KJ/mol

Where I_1, or I_2 or and I_3 is of first, second and third ionization energy of aluminum atom in gaseous state respectively.

The first ionization energy increases across a period from left to right. As we go from left to right in the periodic table, electron is filled in the same principal quantum number and proton is filled in the nucleus. Therefore, attraction between outermost electrons and nucleus increases. This is the reason for increasing first ionization energy when we go along period from left to right.

On the other hand, first ionization energy decreases in going down a group in the periodic table. In this case, as going down a group, electrons are filled in the higher principal quantum number. Therefore, the distance between nucleus and outermost electron is increased and hence, ionization energy decreases going in a group from top to down.

Electron affinity:

Electron affinity is the energy change associated with the addition of an electron to a gaseous atom.

e.g.
$$X(g) + e^- \longrightarrow X^-(g)$$

Atomic radius:

Atomic radius decreases when going a period from left to right. This is due to increase in attraction between the same outermost electron and nucleus. Atomic radius increases from top to bottom in a group of the periodic table.

Schrodinger time independent equation:

Mathematically, Schrodinger time independent equation is written as

$H\Psi = E\Psi$, where H = Hermition operator, Ψ = wave function and
$\qquad$ E = Eigen value (energy)

When H acts on a wave function, it produces eigen value with integer. Therefore, each energy state has its own wave function (Ψ) and energy.

Particle in a box:

Principle of a particle in box is suited for several real systems of physical interest and these are translational motion of ideal gas molecules, electrons in metals and pi electrong in conjugated hydrocarbon and related molecules. Recently, it explains blue shift observed in semiconductor particles. The particle in a box forms the following set of rules, $E = n^2h^2/8ma^2$, n = 1, 2, 3 ….

1. With increasing energy levels, the separation between two consecutive energy levels increases. This is unlike energy levels

formed for hydrogen atom. The energy levels of a particle in one-dimensional box are shown in the following Fig.9.

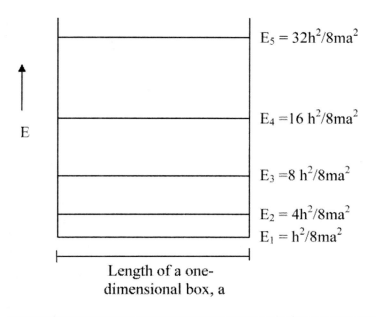

Fig.9: Energy level diagram for a particle in 1-D box

2. At the lowest level, probability of finding the particle is maximum at the centre of the box and this contradicts to the classical mechanics. With increasing energy levels, the probability of finding the particle is more towards walls. Thus, in the higher energy levels the particle in the box resembles classical mechanics.

Harmonic oscillator:

In a simple harmonic oscillator energy levels are equally spaced. The mathematic expression for energy levels of simple harmonic oscillator is given as

$E_n = (v+1/2)h\upsilon$, where $v = 0, 1, 2, 3, \ldots\ldots$

Even when v = 0, $E_0 = 1/2h\upsilon_0$. Energy is not equal to zero. Therefore, this energy level is called zero point energy. The energy levels of a simple harmonic oscillator are given below in Fig.10.

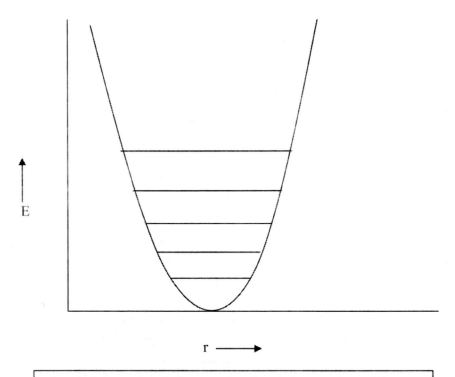

Fig.10: Energy level diagram for simple harmonic oscillator.

In the real system, nonharmonic oscillator is observed. Therefore, the energy levels of nonharmonic oscillator will be different from the energy levels of harmonic oscillator. Thus, with increasing energy, the energy levels of nonharmonic oscillator are compressed and energy level diagram of nanoharmonic oscillator is given in Fig.11.

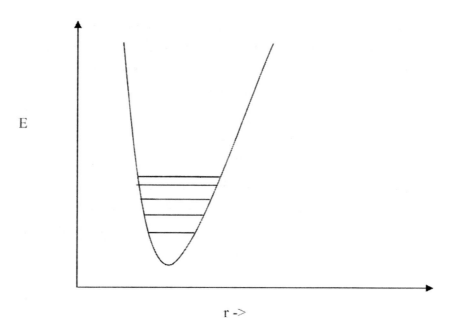

Fig.11: Energy levels in non-harmonic oscillator

Rigid Rotor:

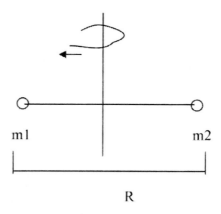

Fig.12: Representation of rigit rotor

Consider two different masses, m1 and m2 rotate with a fixed axis (Fig.12). The distance between the two masses is R and it is also fixed. The calculated energy equation for the rigid rotor is

$E = J(J+1)h2/2I$, $J = 0,1,2,3,$---

The calculated energy levels indicate quantized and energy gap between two consecutive energy levels is increased with increasing energy level. But, at the lowest energy level, the rigid rotator becomes at rest. i.e. it has zero value. That is it does not rotate at zero value of J. Allowed transitions in rigid rotor are explained by electric dipole mechanism. Thus, they are observed between two consecutive energy levels, $\Delta J = 1$.

CHEMISTRY OF SULFUR

Compounds of sulfur as one element are different from that of oxygen element even though they are in the same group in the periodic table. This main difference of formation of number of bonds of sulfur and oxygen with other elements could be understood from their electronic configurations and availability of outermost orbitals.

The electronic configuration of oxygen is $2s^2\ 2p^4$ in the outermost orbitals. In the second principal quantum numbers, there are 2s and 2p orbitals only available. Therefore, oxygen can share two unpaired electrons with other element to form two covalent bonds.

The electronic configuration of sulfur is $3s^2\ 3p^4\ 3d^0$ in the outermost orbitals. In the third principal quantum number, there are three possible orbitals such as 3s, 3p and 3d available. Therefore, the number of unpaired electrons in sulfur atom can be increased by transferring the 3s and 3p paired electrons to the available 3d orbitals. Thus, sulfur atom can have four and six unpaired electrons to form covalent bonds in addition to two unpaired electrons like in found in oxygen atom. The increase in number of unpaired electrons in sulfur is explained below.

The ground state electronic configuration of sulfur is $3s^2\ 3p^4\ sd^0$. 1 electron from 3p orbital is promoted to 3d orbital to the formation of four unpaired electrons ($3s^2\ 3p^3\ 3d^1$). These four unpaired electrons can involve in forming four bonds with other elements. Similarly, promotion of 1 electron from 3s orbital and one electron from 3p orbital to 3d orbitals leads to formation of six unpaired electrons and these six unpaired electrons can involve in the formation of six bonds with other elements.

Oxides and Oxyacids of sulfur:

Sulfur can form compounds with oxygen. There are the two most important compounds of sulfur with oxygen known and these are SO_2 and SO_3. It is important to mention here a main difference between oxygen and sulfur even though they are in the same group of periodic table. Oxygen can combine with another oxygen to form the most stable O_2 compound and the O_2 is a gas at RT and 1 atm. Pressure. Similarly, can sulfur form a compound with oxygen to get SO? Also, S_2 does not exit at RT. At temperature between 800° and 1400°C, sulfur vapor consists mainly of diatomic moeclues (S_2).

Research needed:

My idea at this stage is that SO molecule can be stabilized with metal ions. Therefore, an immense research activity is required to trap SO molecule using transition or main group or rare-earth metal ions.

Similar to O_2, S_2 does not exist at RT and 1 atm pressure. Another direction in the topic of sulfur is that can S_2 molecule be stabilized with metal ions?.

Synthesis of Sulfur Dioxide:

One of the well-known oxides is SO_2. Its synthesis, structure, properties and uses are explained below in addition to sulfurous acid.

Synthesis of SO_2:

A commonly used method for the synthesis of SO_2 involves burning sulfur in air or O_2. About 6-8% of sulfur is transformed into SO_3 as a side product in the burning of sulfur.

$S + O_2 \rightarrow SO_2$ (Main Reaction)

$2S + 3O_2 \rightarrow 2SO_3$ (Side Reaction)

Another method of synthesis of SO_2 is heating of metal sulfide. For example, iron sulfide (FeS_2) is heated in oxygen atmosphere to get SO_2 gaseous product and solid Fe_2O_3. The reaction is represented below.

$$4FeS_2 + 11O_2 \rightarrow 2Fe_2O_3 + 8SO_2$$

Laboratory method of synthesis of SO_2 is the reduction of sulfuric acid using sulfur or copper or mercury. The following equation is represented to explain the redox reaction to get SO_2 using copper and H_2SO_4.

$$Cu + 2H_2SO_4 \rightarrow CuSO_4 + 2H_2O + SO_2\uparrow$$

Structure of SO_2:

The structure of SO_2 resembles with that of O_3. Oxygen can have one sigma bond and one pi bond with sulfur. Therefore, sulfur can have two sigma bonds with two different oxygen element. One pi bond is formed for three elements (O, S & O). The bond angle for O-S-O is 120° and it is planar. Therefore, it involves sp^2 hybridization of sulfur.

Properties and uses of SO_2:

1. SO_2 dissolves in H_2O to yield H_2SO_3. H_2SO_3 is called as sulfurous acid whereas SO_2 is called sulfurous anhydrate. It is stable only in water and in H_2O, sulfurous acid dissociates as given below.

   $$SO_2 + H_2O \rightarrow H_2SO_3 \leftrightarrow 2H^+ + SO_3^{2-}$$

2. Sulfur is hexavalent element and in SO_2, there are only four bonds satisfied. Therefore, SO_2 is an unsaturated compound and it involves in addition reactions with O_2, Cl_2, PbO_2 or BaO_2.

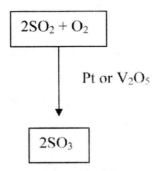

This reaction of addition of O_2 onto SO_2 is catalyzed by V_2O_5. This means that V_2O_5 is not affected by SO_2. Similarly, Cl_2 and PbO_2 can be added to SO_2.

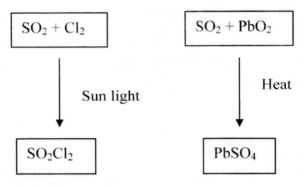

3. Uses of reduction property of SO_2:

   Reduction property of SO_2 is explored in food to reduce vegetable coloring matter and on standing in air, the colorless reduction product is re-oxidized by atmospheric oxygen to the original color. Therefore, the reduction property of SO_2 is temporary.

4. Acid Raid:

   SO_2 is readily oxidized into SO_3 in the atmosphere which is then reacted to H_2O to form H_2SO_4. Then, H_2SO_4 will fall with rain to the earth and hence, this type of rain is called acid rain. Therefore, it is essential to prevent the SO_2 gas contaminating the earth atmosphere.

5. Antioxidant:

   SO_2 can be used as antioxidant to preserve fruit drinks and wines.

Sulfur Trioxide:

Preparation of SO_3:

It is usually synthesized by heating ferrous sulphite and the equation below represents the formation of gaseous SO_3 from $Fe_2(SO_4)_3$.

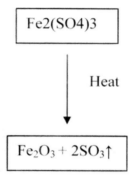

Structure of SO_3:

Gaseous Structure:

This is a planar structure. Oxygen element can have one sigma bond and one pi bond with sulfur. Since sulfur is a hexavalent element, it has three sigma bonds and three pi bonds with three different oxygens. Thus, the planar structure of SO3 is represented below (Fig.13). The angle of O-S-O is 120o and hence sulfur has sp2 hybridization.

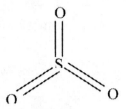

Fig.13: Gaseous Structure of SO₃

Solid State Structure of SO$_3$:

Solid State Structure of SO$_3$ can be a long chain or cyclic one. These two structures are presented below one by one in Figs. 14 & 15.

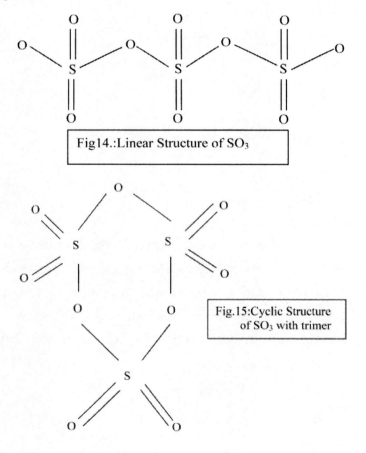

Fig14.: Linear Structure of SO₃

Fig.15: Cyclic Structure of SO₃ with trimer

Important property of SO$_3$:

It reacts with water to form H$_2$SO$_4$, sulfuric acid as shown below.

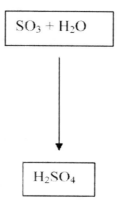

Sulfuric acid is used in a lot and one of the important applications of it in organic synthesis is that it acts as dehydrating agent.

Research Needed in SO$_3$:

Central element in SO$_3$ is sulfur and its ground state electronic configuration is 3s^2 3p^4 3d^0. Sulfur has three sigma bonds and three pi bonds with three different oxygens. Therefore, electronic configuration of sulfur while bonding with oxygen in SO$_3$ is 3s^1 3p^3 3d^2. One s orbital and two p orbitals hybridize to form three sp^2 orbitals and that they form three sigma bondings with three different oxygens. Remaining one p electron and two d orbitals involve three pi bondings with oxygen.

My thought in the SO$_3$ is the exploration of three empty 3d orbitals. Therefore, my question is that can SO$_3$ accommodate 3 pair of electrons to fill the three empty 3d orbitals? Thus, is it right to direct a research activity to synthesize (SO$_3$)$_1$Ca$_3$ compound?

In the proposed compound, (SO$_3$)$_1$Ca$_3$, the 6 electrons from 3 calcium element donate to SO$_3$ to fill the empty 3d orbitals completely. Therefore, there is an electrostatic interaction between SO$_3$ anion and calcium cation to form an ionic compound, (SO$_3$)$_1$Ca$_3$. Thus, the electrostatic interaction can be a driving force for the synthesis of $(SO_3)_1Ca_3$.

BORANES AND RELATED COMPOUNDS

All the chemical bondings are described by two-center-two-electron bonds. This theory of bonding explains mostly all the chemical bondings. For example, a two-center-two-electron bond is explained by considering the formation of covalent bond between two hydrogen atoms in H_2 molecule.

H XX H   Where H represents electron

Each hydrogen atom has one 1s electron and each hydrogen atom shares its outermost electron to form a stable covalent and sigma bond. Thus, the stable covalent and sigma bond consists of two electrons. The two-center-two-electron bond is satisfied. But, this theory can not be extended into boranes even though stable boranes were prepared by German scientist Alfred Stock for the first time. Therefore, boranes form a special compounds from the point of view of bondings. A typical example of boranes is B_2H_6. This compound is conventionally called diborane. The structure of the diborane is given below (Fig.16).

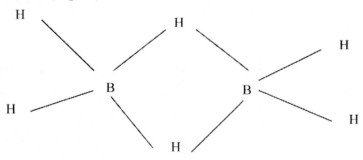

Fig.16: Structure of B_2H_6

The structure of diborane reveals two types of bondings. One type, as found in terminal B-H bonding, is two-center-two-electron covalent and sigma bonding as observed in normal bonding. These two electrons are localized between B and H atoms. The other bonding, observed in the bridging of B-H-B, is three-center-two-electron bonding. This is the delocalized sigma bonding among three atoms (two borons and one hydrogen). Thus, this bonding does not satisfy by normal theory that explains almost all the bondings. Experiments show that the special bridging bonding has longer bond length. The special bonding is called banana bonding. Thus, boranes are electron deficient compounds and hence, they are reactive in nature and form adducts with electron donor moiety. Very similar to boranes, aluminum also forms aluminum-hydrogen compounds. For example, Al_2H_6 has structure very similar to B_2H_6 in the gaseous state.

Some borons in boranes are partly replaced by carbon and these compounds are called carboranes. Boranes and carboranes are neutral or anionic compounds and they are classified into four groups. These four groups are closo [cage, completely closed cluster of n boron atoms], nido [nest, nonclosed Bn-1 clusters], arachno [web, Bn-2 clusters] and conjuncto [joined, formed by linking two or more of the above]. A typical example for carboranes is the charge neutral $C_2B_{10}H_{12}$. This compound is the mostly studied one. This is a solid at room temperature and has melting point 320°C.

Research needed:

When B in $B_{12}H_{12}$ can be replaced partly by carbon to get stable $C_2B_{10}H_{12}$ solid, is it possible to replace partly boron in $B_{12}H_{12}$ by aluminum since aluminum is the group of boron atom to get $B_{12-x}Al_xH_{12}$. Therefore, it is interesting to synthesize and study chemistry of this type of new compounds. Literature survey indicates that Al_2H_6

does exit and hence, it is possible, according to me, to synthesize $B_{12-x}Al_xH_{12}$ successfully.

KETO-ENOL TAUTOMERISM

Keto and enol forms are isomers, and aliphatic ketones exist largely in keto isomer with a trace amount of enol isomer at equilibrium in solution. For example, beta-diketones and beta-diketonates having keto-enol tautomerism are well studied and these diketones in the enol form exhibit intramolecular hydrogen bonding. Thus, intramolecular hydrogen bonding favors enol form stable in the solution. The following illustration describes the keto-enol tautomerism of diketons (Fig.17).

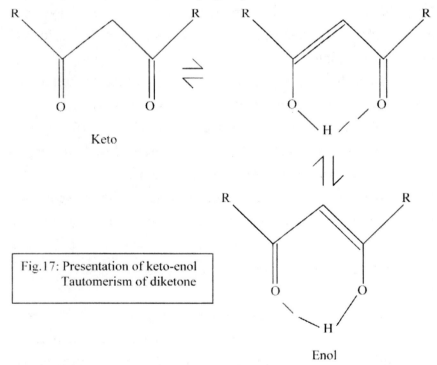

Fig.17: Presentation of keto-enol Tautomerism of diketone

It is also possible to stabilize a particular isomer either in keto or in enol form in a large amount in solution by adjusting pH of the solution. Thus, acidic condition favors keto isomer and basic condition allows enol isomer in large quantity.

One important question arises here that what is the tool available now to study the keto-enol tautomerism. The best tool is solution and solid nuclear magnetic resonace (NMR). Some atoms have odd neculear spin and these atoms have nuclear spin energy levels. These levels are degenerate in the absence of magnetic field. When magnetic field is applied on these atoms, the degeneracy is lost and depending upon the magnetic field strength, the energy separation of nuclear spin energy levels varies. The basic principle underlying this concept of spin energy levels is presented below in Fig.18.

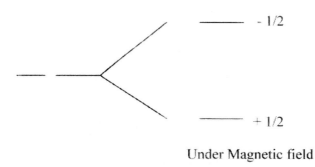

Under Magnetic field

Fig.18: Nuclear spin energy levels with and without magnetic field

The energy separation under magnetic field lies in the region of radio frequency. Therefore, under fixed radio frequency, magnetic field is slowly varied to get resonance with radio frequency. Under the resonance condition nuclear spin is flipped and the depending upon the abundant of a particular nuclues the integration of resonance peak varies. Thus, nuclear magnetic resonance (NMR) can be successfully explored to find out of a particular atom in the solution or in the solid state. This is also important observation that resonance peak of atom largely depends upon the surrounding of the atom.

Research needed:

So far, keto-enol tautomerism is described in diketones. My thought is to extend the keto-enol tautomerism in aliphatic or aromatic dicarboxylate dihydrazides. The tautomerism, for example, in oxalyl dihydrazide (ODH) can be represented below in Fig.19.

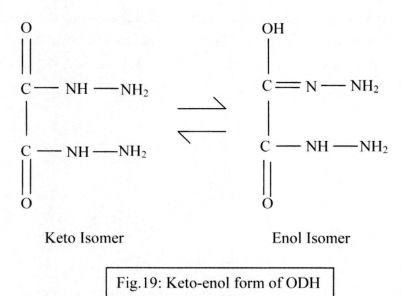

Fig.19: Keto-enol form of ODH

Under different pH of the solution, different isomer can be stabilized. The next step is to allow a particular isomer to react with main group or transition or rare-earth metal ions to precipitate a particular isomer as a complex compound. Once this is achieved, NMR can be used to characterize powder sample. Because of different environment around carbon and nitrogen, NMR is a right tool to do N and C NMR.

The next step is to get single crystal of a particular isomer with metal ion as a complex compound. This can be achieved using hydro or solvo thermal method as described in the inorganic materials synthesis chapter.

Another important question dealt with ligand isomer is the coordinating atoms of ligand to metal ion in the complex compound. There are atoms of ligand ready to participate. Therefore, a linear or a chelating behavior of ligand isomer with metal ions can be isolated. Chelating complex compounds are useful to store hydrogen molecule within the pores formed in the chelating structure. Therefore, the following outcomes can be expected from keto-enol tautomerism.

1. Stabilization of a particular isomer of hydrazides in keto-enol tautomerism.
2. Complexing a particular isomer of hydrazides with metal ions in a solid state.
3. This tautomerism helps to study solution or solid state NMR of carbon and nitrogen.
4. Knowledge in single crystal group is required to isolate complexes.
5. Interesting structural features can be studied.
6. Physisorption of hydrogen molecule using metal complexes of chelating ligands engenders new direction in the hydrogen economy.

COORDINATION COMPLEXES

In order to understand functions of Vitamin B12, hemoglobin, chlorophyll and others, it is very important to study the structural chemistry of them. These compounds are complex materials. Therefore, in this topic, I try to explain chemistry of complex materials. Thus, structural chemistry of coordination complexes is essential to understand the properties of them. Or in other words, properties of complex materials are depended upon their structures of them.

Coordination Compounds:

After Alfred Werner, a German-Swiss chemist provided reasonable theory to explain coordination compounds, the most of the inorganic chemists directed their research towards coordination compounds in the 20th century. To understand the theory, representation and bonding of coordination compounds, let us consider a classical coordination compound, cobalt (III) chloride ammonia. The nominal composition of the coordination compound is given as $CoCl_3.6NH_3$.

There are two types of bonding of chloride anion and ammonia neutral molecule to cobalt ion respectively. Thus, ammonia donates its nonbonding electrons to cobalt ion and this type of bonding is named as coordination bonding or secondary valence bonding whereas chloride anion has ionic bonding with cobalt atom and this is called primary bonding. Six ammonia molecules coordinate to cobalt (III) cation by an octahedral environment and it is represented below (Fig.20) (Sidgwick's Theory).

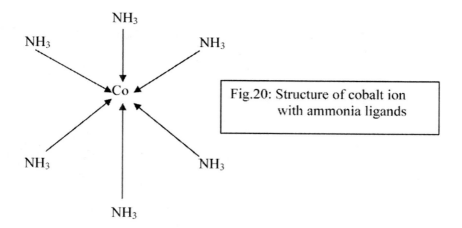

Fig.20: Structure of cobalt ion with ammonia ligands

The representation clearly indicates possibility of variety of cation structures with different metal and ligands. Another well known structure in the cobalt chloride-ammonia coordination compound is tetrahedral coordination and this is represented below (Fig.21).

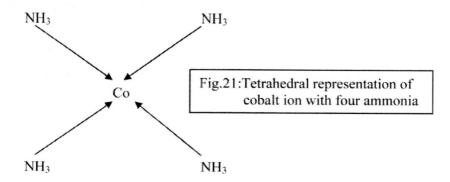

Fig.21: Tetrahedral representation of cobalt ion with four ammonia

The ligand, ammonia molecule is explored to explain its coordination behavior with cobalt cation. It is interesting to note that the ligand is a neutral molecule. Another type of ligands is anion. The typical example for anion ligand is

$[Co(NH_3)_4Cl_2]Cl$

Here, four neutral ammonia molecule and two chloride anion are coordinated to cobalt (III) cation. The molecular structure of [Co(NH$_3$)$_4$Cl$_2$]Cl is represented below (Fig.22).

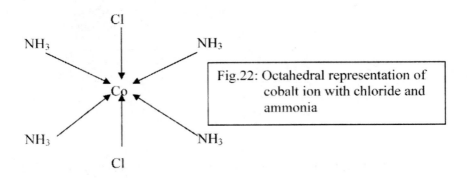

Fig.22: Octahedral representation of cobalt ion with chloride and ammonia

So far, we have seen two types of ligands. One is a neutral molecule and other one is anion ligand. Another interesting ligand is hydrazinium monocation. There are two nitrogen atoms present in hydrazine molecule. One of them is positive cation and another is ready for coordinating with central metal ion like ammonia. Therefore, as a whole, hydrazinium monocation is positive charge, it can act as a ligand.

Structure and function of hemoglobin:

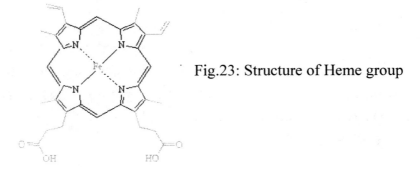

Fig.23: Structure of Heme group

Hemoglobin is an iron containing coordination compound and it involves oxygen transportation in red blood cell. It is an assembly of four protein subunits. The non-protein heme group is important to mention in this section since it has complex coordination structure as shown above in the figure 23. Thus, four heterocyclic rings containing nitrogen (coordinating atom) are coordinated to central iron ion. The fifth one is bound strongly to protein. Therefore, a sixth position is available for oxygen transportation. When oxygen is absent to fill the sixth position of iron atom, water molecule weakly coordinates to iron atom.

MATERIALS CHEMISTRY

SYNTHESIS OF INORGANIC MATERIALS

Synthesis chapter in the materials chemistry book is essential since without synthesizing them, no studies can be performed. Therefore, I have taken steps to put all the known and required methods together in this chapter. Thus, this chapter starts with general conventional synthesis of inorganic materials which is then followed by nonconventional methods. The nonconventional methods include solid state metathesis, Igneous combustion, hydro/ solvothermal, ionothermal, ion-exchange and intercalation, reverse micelle or microemulsion, precursor, coprecipitation, microwave, sol-gel and electrochemical methods.

Conventional Solid State Method:

General Concept:

In this method, at least one of the reactants is a solid phase. This method is widely used, even now-a-days, in academia and industries as well. Thermodynamically favorable phases are obtained in this method. However, room temperature is not sufficient enough to convert reactants into thermodynamically stable product. Therefore, high temperature is essential to convert reactants into stable product. This method of synthesis of inorganic materials involves heating the corresponding component of oxides or carbonates in a stoichiometric ratio at a very high temperature in an inert container such as quartz or alumina or platinum. For example, oxide material synthesis requires temperature in the range 500-2000°C.

This is a diffusion control reaction because of solid state reaction. At this stage, solid state reaction differs from organic and organometallic synthesis. Therefore, high temperature heating is required. Fig.24 illustrates a solid state reaction. Thus, reactants, A & B are ground well and the physical mixture is heated at a particular temperature. Then, repeated grinding and heating are performed to make sure all the reactants are converted into product.

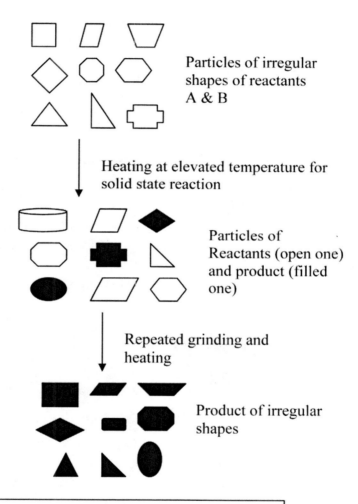

Fig. 24: Representation of a solid state reaction

Once product is formed, it functions as a barrier for further reaction of reactants into product. Also, with course of solid state reaction, thickness of product increases and therefore, diffusion distance does too. Therefore, repeated grinding is required to make contact, for example, between two reactants so that they will react. Barrier concept is explained pictorially in the following reaction (Fig.25).

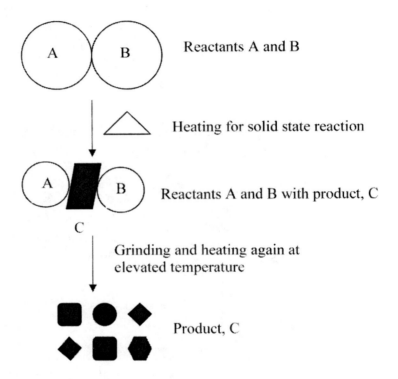

Fig.25: Application of repeated grinding and heating in a solid state reaction

The most advantage of solid state technique is that all the reactants are converted into solid product and hence, the solid state reaction is called pollution free reaction. A disadvantage of solid state technique is the particle size limitation. i.e. even after mechanical grinding the particle size is at most achieved to sub-micron. Therefore, wet-chemical methods are useful in this regard.

Starting materials employed in the solid state technique for oxide synthesis are oxides or carbonates or hydroxides. For an effective reaction, starting materials should be particle size as low as possible. Therefore, carbonates as starting materials are found useful since they lead to formation of reactive oxides 'in situ' in this type reaction.

Experimental Procedure:

This is explained by considering solid state reaction between $SrCO_3$ and SiO_2 to get Sr_2SiO_4. Usually, pre-heat treatment well below the formation temperature is required as a warm-up treatment. For this reaction, the warm-up treatment is 600°C. Then, it is heated with temperature as high as 1300°C to get the required phase. Some time, post-treatment is needed for critical compounds such as high-temperature superconductor, $YBa_2Cu_3O_{6.9}$. In order to accelerate solid state reaction flux is added with reactants. Flux method explores low melting materials as a component in the reaction mixture of solid state reaction. At the solid state reaction temperature, the flux material melts and one of the reactants is soluble in the flux liquid. Therefore, solid state reaction is accelerated. After the reaction is completed, flux material is washed out with solvent. The main requirement of a flux material is that flux material should not react with reactants and product to get unwanted product.

In a typical experiment, 50g $SrCO_3$, 13.04g $BaCO_3$, 12.4g SiO_2, 1.73g EuF_3 and flux, 1.2g NH_4Cl are mixed well to get the nominal composition of $(Sr_{1.64}Ba_{0.32}Eu_{0.04})SiO_4$. This mixture is ball milled in presence of small amount of methanol at room temperature for 24 h. Then, this mixture is fired at 600°C for 2 h. Then, it is ground well and the final heat-treatment at 1300°C is carried out in the fllow of mixture of H_2 and N_2 gases. Formation of host and presence of activator phosphor are confirmed by powder XRD and fluorescence spectra respectively. Emission spectrum obtained by exciting at 450 nm shows a broad band peaking at 560 nm. Fig26 shows emission and absorption spectra of Eu^{2+} activated strontium orthosilicate.

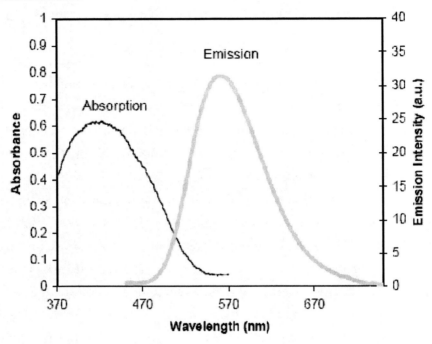

Fig.26: Absorption and emission spectra of Eu^{2+} in orthosilicate

Non-conventional Methods:

Non-conventional techniques are defined as those techniques which do not comprise of the normal mixing, calcinations and grinding operations. Non-conventional methods include solution and vapor-phase processing techniques and thermal decompositions of metal compounds. Solution or chemical processing techniques are classified according to the means of removing the solvent. These techniques lead to formation of homogenous constituents and hence, repeated heating and grinding can be avoided. Thus, non-conventional methods include co-precipitation, sol-gel, hydrothermal, solid state metathesis and igneous combustion to name a few. Some of non-conventional methods are described below one by one.

Solid State Metathesis Method (SSM):

General Concept:

Dr. Kaner (USA) and Dr. Parkin (UK) independently reported the synthesis of a variety of materials by solid state metathesis (exchange) reactions. This method involves exchange exothermic reaction between reactive metal halide with alkali metal main group compound. The driving force for this type of reaction is the formation of thermodynamically stable salt such as AX where A=Na or K & X=Cl, Br, I. Prime importances of this type of reaction are

i. preparation of anion solid solutions
ii. Obtaining various particle sizes of products.

A serious limitation of this procedure is the requirement of anhydrous halides which require handling of reactants in dry box and storage in presence of inert atmosphere.

Experimental Procedure:

Technologically important materials such as superconductors, semiconductors, magnetic materials, intermetallics and so on are synthesized by this method. For example, the synthesis of ZrO_2 is explained here. ZrO_2 has three structural modifications. One is monoclinic zirconia (room temperature form), the second is tetragonal zirconia and the third is cubic zirconia (high temperature forms). The phase formation depends upon temperature. However, t- or c- zirconia could be stabilized at room temperature by doping of Y_2O_3, CeO_2, MgO or CaO at Zr site of ZrO_2. The reactants employed in this method for the synthesis of zirconia are anhydrous $ZrCl_4$ and Na_2O. Once igniting the starting mixture of $ZrCl_4$ and

Na_2O, this reaction becomes rapidly self-sustaining and can reach high temperature within a short period. The formation of zirconia is represented by the following equation.

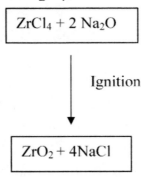

Thus, the reactant mixture consists of 1.0g $ZrCl_4$, 0.532g Na_2O. Na_2O_2 could also be used instead of Na_2O. These precursors are ground in an inert atmosphere. Then, it is ignited using heated nichrome filament in an argon-filled glove box. After a few minutes, reaction gets over. Then, the product is washed with ethanol, followed by 2M HCl to dissolve an unreacted starting material. This reaction leads to formation of monoclinic zirconia. Tetragonal or cubic zirconia could be obtained by adding 9 atom% CaO or Y_2O_3 or CeO_2. Now-a-days, this reaction is extended to synthesis complex oxides by Prof. J. Gopalakrishnan, India.

Both the cationic and anionic solid solutions could be synthesized by SSM. Thus, the SSM has been extended to prepare $(Mo, W)S_2$ from $MoCl_5$ and WCl_3 precursor, and $Mo(S,Se)_2$ is also prepared by a reaction among $MoCl_5$, Na_2S and Na_2Se.

Research needed:

Therefore, I strongly believe that anion solid solution such as Zn(O,S) could also be prepared by the SSM. Because, ZnS is a photocatalyst for H_2 evolution from water and ZnO is good for O_2 evolution from water. Combination (solid solution) of these two catalysts might be good photocatalyst for simultaneous evolution of H_2 and O_2 from water.

Igneous Combustion Method:

General Concept:

This method of synthesis of oxides is new and hence, it is explained in detail. Combustion process is an exothermic reaction, which occurs with evolution of light and heat. This is usually represented by a triangle (Fig.27) and each corner represents oxidizer, fuel and ignition.

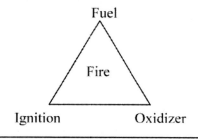

Fig.27: Explanation of combustion

Therefore, it is important to ignite the mixture containing appropriate amounts of fuel and oxidizer. Combustion is simply expressed by a well-known reaction, burning of carbon in presence of oxygen.

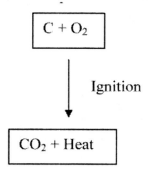

Here, carbon is reducer and O_2 is oxidizer. For the combustion synthesis of oxides, carbon is replaced by urea or glycine or hydrazine-based compounds. O_2 is replaced by metal nitrates, ammonium nitrate and ammonium perchlorate. Therefore, usually, oxidizer contains metal ion. But, sometime, fuel such as ammonium meta vanadate contains metal ion. Boric acid can be used as a neutral source for boron in the synthesis of borates.

Stoichiometric compositions of metal nitrates and fuel are calculated based upon propellant chemistry. Thus, heat of combustion is maximum for O/F ratio unity. Based on the concepts used in propellant chemistry, the elements C, H, V, B, or any other metal are considered as reducing elements with valences 4+, 1+, 5+, 3+ (or valence of the metal ion in that compound), respectively and oxygen is an oxidizer having the valence of 2-. The valence of nitrogen is taken as zero because of its conversion to molecular nitrogen during combustion.

The oxidizing and reducing valences of metal nitrates and fuels used in the combustion synthesis of oxide phosphors are summarized in table 2.

Table 2: Valences of metal nitrates and fuels

Compound	Oxidizing or reducing valence
$M(NO_3)_2$	10-
$M(NO_3)_3$	15-
$M(NO_3)_4$	20-
NH_4NO_3	2-
Urea, CH_4N_2O	6+
Oxalyl Dihydrazide, ODH, $C_2H_6N_4O_2$	10+
Carbohydrazide, CH, $CH_6N_4O_2$	8+
3-methly-pyrazole-5-one, 3MP5O, $C_4H_6N_2O$	20+
Diformyl Hydrazine DFH, $C_2H_4N_2O_2$	8+
NH_4VO_3	3+
H_3BO_3	0

Calculation of molar ratio of reactants:

Y_2O_3 synthesis using $Y(NO_3)_3$ and urea is described in this section. Thus, for complete combustion of 2 moles of $Y(NO_3)_3$, 5 moles of urea is required.

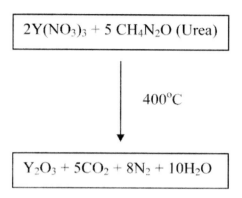

Experimental Procedure:

Combustion method is yet another wet-chemical method which does not require calcination and repeated heating. During combustion process lots of heat evolve. Such a high temperature leads to formation and crystallization of oxides.

Wet-chemical techniques are now available for simple oxide phosphor such as Eu^{3+} activated Y_2O_3 red phosphor. These wet-chemical methods dope rare-earth activators uniformly. But, calcination is required to get crystalline (required phase) phosphors. However, wet-chemical techniques are not available for the synthesis of complex oxide phosphor (green lamp phosphor).

Tb^{3+} activated $(La,Ce)MgAl_{11}O_{19}$ green phosphor is obtained by rapidly heating an aqueous concentrated solution containing stoichiometric amounts of metal nitrates [$La(NO_3)_3$, $Ce(NO_3)_3$, $Tb(NO_3)_3$, $Mg(NO_3)_2$] and urea at 500°C. Thus, $M(NO_3)_2$:urea (1:1.66), and $M(NO_3)_3$:urea (1:2.5) redox compositions are used for the combustion synthesis of (La,Ce)MgAl$_{11}$O$_{19}$:Tb^{3+}. Theoretical equation assuming complete combustion can be written as follows:

$$1-x-y\,La(NO_3)_3 + xCe(NO_3)_3 + yTb(NO_3)_3 + Mg(NO_3)_2 + 11Al(NO_3)_3 + 31.67\ CH_4N_2O\ (urea)$$

Ignition Temperature = 500°C

$$La_{1-x-y}Ce_xTb_yMgAl_{11}O_{19} + by\ products$$

Powder XRD pattern of green phosphor reveals single phase crystalline in nature. This observation is notable because ceramic synthesis of green phosphor requires elevated temperature (>1400°C) and always contains Al$_2$O$_3$ as impurity.

Emission spectrum of combustion synthesized LaMgAl$_{11}$O$_{19}$:Ce^{3+} shows a broad band at 340 nm. The emission of Ce^{3+} is due to 4f^{6}5d → 2F$_j$ transition. Addition of Tb^{3+} in LaMgAl$_{11}$O$_{19}$:Ce^{3+} results in emissions at 480 and 543 nm in addition to Ce^{3+} emission. The Tb^{3+} emissions which arise due to energy transfer from Ce^{3+} are attributed to $^5D_3 \rightarrow {^7F_5}$ and $^5D_4 \rightarrow {^7F_5}$ transitions. Thus, presence or doping of sensitizer (Ce^{3+}) and activator (Tb^{3+}) is confirmed by fluorescence spectra (Fig.28).

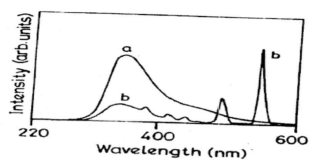

Fig.28: Emission spectra of Ce^{3+} alone (a) and Ce^{3+} and Tb^{3+} (b).

Research Needed:

$CaWO_4$ scheelite compound can be converted into $AA'(MO_4)_2$ type system where A = alkali metal ion, A' = Rare earth ion and M = Mo or W or both. The scheelite system has been explored as a good host for Eu^{3+} ion doping to get red phosphor that is useful in solid state lighting.

My idea in this direction is to replace calcium ion in $CaAl_2O_4$ spinel by alkali ion and rare-earth ion to get aluminate based phosphor. Since Al_2O_3 is a refractory material, it requires more than 1000°C to synthesize $A_{0.5}A'_{0.5}Al_2O_4$. At such high temperature, alkali metal ion might be evaporated, which leads to inhomogeneous product. Therefore, igneous combustion synthesis, which is low temperature initiated but exothermic reaction, can be successfully explored to synthesize $A_{0.5}A'_{0.5}Al_2O_4$ by choosing a suitable fuel to get aluminate based phosphor.

Hydrothermal Method:

General Concept:

Theoretically speaking, hydro/solvo thermal technique could be employed for three major applications such as single-crystal growth, nano materials synthesis and recovery of metals from ores.

This method involves heating the corresponding constitutents in water/solvent to its above boiling point in a closed system with autogeneous pressure.

Now-a-days this method is extensively explored to synthesize inorganic materials such as open framework zeoletic materials, nano crystals of metallic powders, nano size oxides of simple and mixed compounds, quantum dots and so on.

Learn and lead methods direct to isolation of single crystals of open framework materials when the reactants deviate from product stoichiometry. Once structure of open framework is solved and product composition is determined, stoichiometric reaction is carried out to get single phase polycrystalline powder.

Reactions are designed to carry out in water or solvent in presence of organic templates. Therefore, the product mostly contains water and hydroxyl group. Organic template directs the reaction to form an open framework structure with 1- or 2- or 3- dimensional pores.

Experimental Procedure:

This method involves reaction between reactants in a closed system in the temperature range $100°$-$250°C$. This method is known as low temperature hydro/solvo thermal method. Most of the materials are precipitated in water medium (known as hydrothermal method). When organic solvents are used in the place of water it is termed as solvothermal method. The container useful for the above mentioned reactions is usually made up of Teflon-lined stainless steel. It is commonly known as autoclave or bomb. Fig. 29 shows schematic diagram of autoclave.

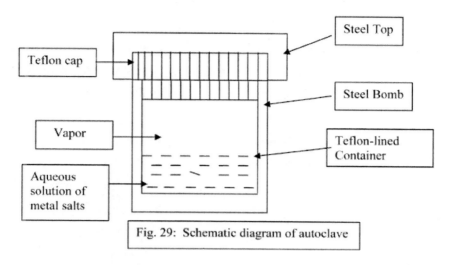

Fig. 29: Schematic diagram of autoclave

Now, scientists couple various reactions with this method and hence various names are required to describe them. These various methods are summarized below one by one.

Straight forward reactions in hydrothermal method:

The hydrothermal method of synthesizing various inorganic materials is based on straight forward reaction between ions and the oxidation states of metal ions in the product are same as in the reactants. Thus, this method explores stable oxidation states to start with and the reactants are mostly soluble in water at room temperature. At the reaction temperature (above room temperature), the ions of reactants react to yield insoluble materials and hence, after the reaction is over, product is easily separated by filtration. To understand this method, formation of nanosize oxides is explained.

For example, hydrothermal synthesis of nanosize $BaTiO_3$ is described. This material is extensively studied because of its dielectric property. Thus, dipole moment is created when electric field is applied and the dipole moment vanishes once electric field is removed. Starting materials for $BaTiO_3$ synthesis by this method are $BaCl_2$ and $TiCl_4$. After a careful mixing of these water soluble chloride reactants, NaOH is added to this mixture and then, the basic solution of the mixture is hydrothermally treated to get tetragonal $BaTiO_3$. In order to avoid the formation of thermodynamically stable $BaCO_3$ phase, water is heated to remove dissolved CO_2 before adding all the constituents to the mixture.

Oxidation reactions in hydrothermal method:

The oxidation reaction of metal powder in acidic solution, namely phosphoric acid, is sometime coupled with hydrothermal method to synthesize single crystals or powders of metal phosphates with interesting open framework structures. When phosphoric acid is replaced by arsenic acid or boric acid the final products can be metal arsenates or borophosphates respectively. The materials synthesized by this method are not possible by any other methods. Thus, this method has its own merits when it is compared with other methods.

For example, mixed valence titanium compounds with open frameworks are synthesized by oxidation of titanium metal powder under acidic condition using phosphoric acid in presence of diamines. Interestingly, titanium metal powder is getting oxidized to ions and then, the ions react with phosphoric acid to form titanium phosphate. Because of presence of diamines, open framework material is obtained under hydrothermal condition. However, the same reaction can not be performed at RT to get open framework titanium phosphate. The reaction conditions required to get mixed valence titanium phosphate with open framework structure is explained in equation below.

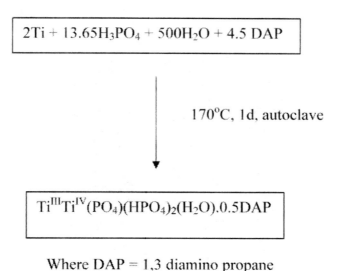

Where DAP = 1,3 diamino propane

In this structure, 1-D pore is seen and in this pore 1,3 diamino propane is occupied. Interestingly, 1,3 diamino propane is not getting oxidized under hydrothermal conditions and this is the reason why it is successful to get open framework phosphate of titanium.

When the same reaction is performed in presence of ethylene diamine (instead of 1,3 diamino propane) another mixed valence titanium phosphate is obtained. This compound has also open framework structure and as in the previous case, this compound can not be synthesized by any other method. The nominal composition of this titanium phosphate is $Ti^{III}Ti^{IV}(HPO_4)_4 \cdot C_2N_2H_9 \cdot H_2O$. The unique in nature of this compound is that the open framework structure remains stable even at 600°C. When this compound is heated at 600°C in air, mono protonated ethylene diamine is expelled out from this compound. For the charge compensation, Ti^{III} is getting oxidized to Ti^{IV} ($Ti_2(HPO_4)_4$). Therefore, the structure is retained without destruction up to 600°C in air. The structure of the compound is shown in Fig. 30.

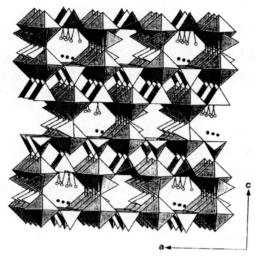

Fig.30: Polyhedral representation of mixed valence titanium phosphate. Ethylene diamine and water occupy in the 1-D pore.

Reduction reactions in hydrothermal method:

In this method, noble metal powder with nanometer in size can be obtained by reducing noble metal ions by ethanol liquid under hydrothermal conditions. To start with this type of reaction under hydrothermal conditions, water soluble noble metal ions are considered. To this aqueous solution fatty acid and ethanol are added. The hydrothermal reaction usually occurs at about 150°C. During hydrothermal reaction metal ions are getting reduced to metal powder by ethanol. These metal powders in the final product is solid and hence, it is easily separated after the reaction is over. Examples for noble metals synthesized in nanometer sizes are Ag, Au, Ru, Rh, Ir, Pt, Pd and son. It is interesting to notice that the added long chain fatty acid keeps the particle size of final products in nanometer size.

Metathesis reaction in benzene thermal synthesis of nano crystalline GaN powder:

Benzene thermal synthesis is a very similar method to hydrothermal method but the main difference is the reaction medium. In this method, benzene is used in the autoclave or bomb instead of water. Therefore, this method is called benzene thermal metathesis synthesis. The following equation describes about this method.

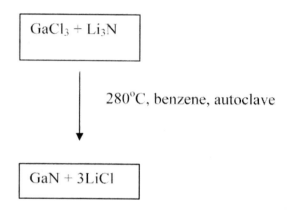

The driving force for this benzene thermal synthesis is also due to thermodynamic controlled reaction of LiCl formation.

In a closed container, the solution based metathesis reaction is carried out at 280°C to get nano powder of GaN. Even though solid state metathesis was famous reaction in the middle 90's in USA at Prof. Kaner group, UCLA, the direct band gap GaN can not be synthesized by the solid state metathesis method. Hence, the combination of metathesis and hydrothermal methods offer a new direction to get nano powder of GaN. When solid state and precursor methods can be used only above 500°C to get GaN and hence, this method is unique method to synthesize GaN powder at lower temperature.

Very recently, Prof. Domen in Japan has explored solid solution of GaN and ZnO as a photocatalyst for water splitting into hydrogen and oxygen gases under visible light. Also, GaN single crystal is being explored to fabricate LEDs.

Research needed:

It is also possible to couple metathesis reaction with hydrothermal method to synthesize novel organically templated or porous open framework materials. Example includes metathesis reaction among sodium borate, sodium silicate and amine or hydrazinium chloride under hydrothermal method.

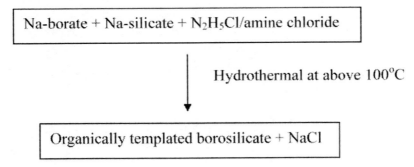

The above mentioned metathesis reaction is thermodynamically driven reaction due to formation of NaCl salt. Another source for silicon is sodium trisilicate, $Na_2Si_3O_7 \cdot xH_2O$. Similarly, borophosphate can also be synthesized by this method starting with sodium borate, sodium phosphate and structure directing groups.

Sol-gel hydrothermal synthesis of mainly titanates:

Hydrothermal method is used for crystallization of oxides that obtained from sol-gel derived amorphous gel. Thus, hydrothermal requires as low as 200°C when it is compared to above 500°C in conventional crystallization of oxides obtained from sol-gel derived amorphous oxides. Also, hydrothermal method leads to formation of nanocrystalline materials whereas conventional method leads to big and agglomerated powder. Because of nanopowder (free from agglomeration) in sol-gel and hydrothermal synthesis method, sol-gel hydrothermal oxides can be sintered very well even at relatively low temperatures.

For example, $(Na_{0.8}K_{0.2})_{0.5}Bi_{0.5}TiO_3$ nanowires can be obtained after gel is obtained by sol-gel method and followed by hydrothermal crystallization at 160°C. These nanowires have large surface area and hence, this nano powder has more reactivity at 1100°C during sintering process. Just two hours are sufficient to get 98% of theoretical density. Sintering process leads to remove pores and makes contact with each grain. Bulk density is the sintered density and materials attain single crystal density. Steps involved in obtaining 98% theoretical density of $(Na_{0.8}K_{0.2})_{0.5}Bi_{0.5}TiO_3$ are shown below in flow chart 1.

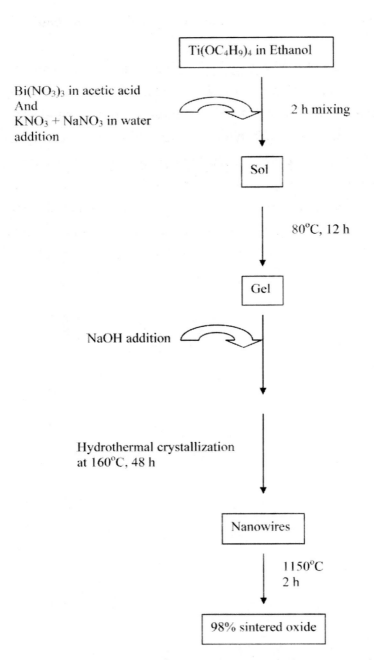

Flow chart 1: Sol-gel at RT, Hydrothermal crystallization at 160°C followed by sintering at 1150°C for $(Na_{0.8}K_{0.2})_{0.5}Bi_{0.5}TiO_3$.

Similar process can be extended to get tetragonal zirconia by combining sol-gel method with solvothermal technique.

Ionothermal Method:

General Concept:

This is a method that makes use of ionic materials as solvent such as 1-methyl-3-ethyl-imidazolium bromide [(emin)]Br. The ionic materials are liquid at the synthesis temperature and also, because of very low vapor pressure of ionic solvent in its melting point, this method definitely differs from hydro/solvothermal method. Therefore, the ionothermal method does not require autoclave or bomb to carry out reaction. Unlike hydro/solvothermal method, this method just requires ordinary round bottom glass container. Atmospheric reaction of the ionothermal method leads to highly safe to carry out reaction and thus, ionothermal method is better than that of hydro/solvo thermal method. Recently, microwave has been explored to accelerate the ionothermal reaction. Thus, this method can be used to synthesize microporous zeolites, metal organic frameworks, nanomaterials, coatings and so on.

Experimental:

In this section, merit of ionothermal method using microwave heating is described very briefly. SAPO-11 is silico aluminophosphate with microporus. Precursor solution for SAPO-11 is made with (emim)Br/Al(OC$_3$H$_7$)$_3$/H$_3$PO$_4$/HF (molar ratio of 32:1:3:0.8 respectively) first after allowing the mixture at 100°C for 4 h in Teflon container. Then, tetraethyl orthosilicate (TEOS) is introduced to the mixture. At this stage, stainless steel substrate is immersed in this solution and entire setup is heated at 150°C for 2 h using microwave. During this step SAPO-11 is coated on the substrate. H$_3$PO$_4$ acid consists of water and hence, presence of little water does not affect the ionothermal method.

Research Needed:

When microwave assisted ionothermal method is good for coating microporous material on alloys, it should be ok for coating of other materials on metal substrate. It is also important that alloy substrate should be immersed completely in precursor solution to avoid microwave reflection by alloy substrate.

My idea in this direction is to coat hydroxy apatite on titanium metal plate using microwave assisted ionothermal method. To get precursor solution, calcium nitrate, phosphoric acid and (emin) Br are required and heat the mixture for a period of time. Then, titanium plate is immersed in this solution and the entire thing is microwaved at a particular temperature. Hydroxy apatite coated titanium plate can be thus obtained by this method. Hydroxy apatite coated titanium plate might find application in the replacement of damaged/diseased bones.

Ion-exchange and Intercalation Methods:

General Concept:

Ion-exchange reaction is a replacement of mostly cations by other cations such that the structure of reactant is retained in the product also. Intercalation is the reaction of interlayer ion with other reactant. Here also, the structure of reactant is retained in the product. These reactions are carried out at low-temperature and hence, called as chemie duce or soft reaction. Such compounds can't be synthesized by high-temperature reactions. Ion-exchange and intercalation reactions are represented in the following equation (Fig.31).

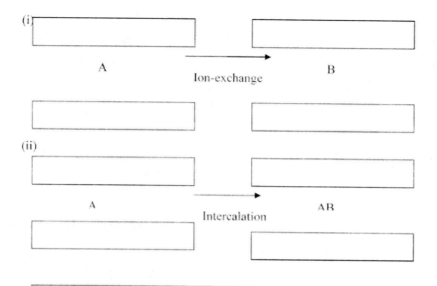

Fig. 31: Schematic diagram of ion-exchange and intercalation reactions.

Experimental Procedure:

(A) Ion-exchange method:

Some cations present in the interlayer of hectorite (Na^+) are replaced by Cd^{2+} ion and hence, this type of reaction is called ion-exchange reaction.

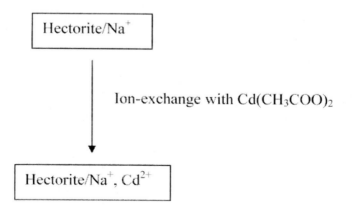

About 10 times excess of $Cd(CH_3COO)_2$ with respect to hectorite weight is dissolved in water. To this clear solution hectorite is added and the solution is allowed for stirring at room temperature for 24 h. Then, this solution is filtered off to get white powder of Cd^{2+} ion-exchanged hectorite (hectorite/Na^+,Cd^{2+}). After ion-exchange reaction, the hectorite obtained is a neutral compound and therefore, some of Na^+ ions are replaced by Cd^{2+} ions. This colorless powder is used for intercalation reaction in the following part.

(B) Intercalation Reaction

In this section, intercalation reaction is described. Intercalation reaction between hectorite/Na^+,Cd^{2+} dispersed in water and H_2S gas (method 1) results in expulsion of CdS from the interlayer of hectorite as represented below.

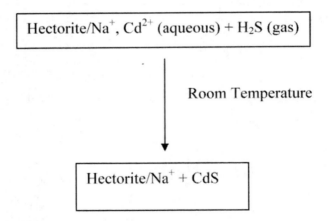

CdS has been incorporated successfully into the inter layer of hectorite by the intercalation between hectorite/Na^+, Cd^{2+} (solid) and H_2S gas (method 2) at room temperature.

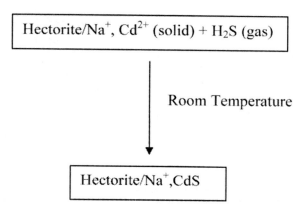

This is an important reaction since one reactant is solid phase and another reactant is gaseous one. Incorporation of CdS in the interlayer of hectorite was confirmed by powder XRD patterns. Hectorite/Na$^+$, CdS prepared by method 2 does not show reflections corresponding to CdS whereas this material prepared by method 1 shows the CdS reflections in the powder XRD pattern. Incorporation of CdS in the interlayer has been further supported by UV-Vis absorption.

Reverse Micelle or Micro emulsion Method
General Concept:

When surfactants in hydrocarbon solvents are mixed with a little quantity of water, inverse micelle or reverse micelle forms. It is made up of spherical shape wherein outer surface is hydrocarbon end of surfactants and inner surface is ionic end of surfactants. Thus, inner surface acts as ionic vessels for ionic reaction. The formation of reverse micelle or inverse micelle is represented below (Fig.32).

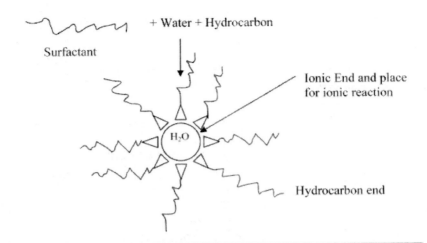

Fig.32: Water microemulsion formation of surfactants in hydrocarbon medium.

Experimental Procedure:

Formation of nanomaterials with various sizes is understood here. The sizes of nanomaterials might be varied by changing the ratio of water/hydrocarbon solvent.

For example, Cd^{2+} ions are dissolved in reverse micelles. At this stage, Cd^{2+} ions occupy in inner sphere of surfactants i.e. in water droplets. Then, H_2S gas is passed through the reverse micelle as a source for S. Since H_2S is an ionic gas, it is getting touch with Cd^{2+} ions. CdS is formed in water droplets. Thus, CdS of nanometer size could be prepared by reverse micelle method. The following diagram with an equation describes the formation of CdS in the reverse micelles (Fig.33).

$$\text{Micro-emulsion} + Cd^{2+} + S^{2-}$$

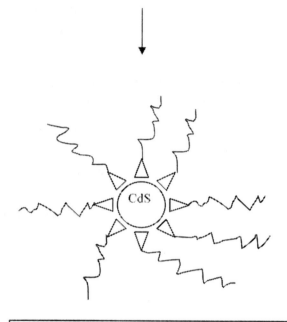

Fig.33: CdS formation in microemulsion

Precursor Method:

General Concept:

Importance of precursor method for the synthesis of inorganic materials is clearly understood by considering the formation of cation solid solution. Formation of cation solid solution is evidenced from similar powder XRD patterns of individual cation compound. When cation solid solution is rapidly heated above its decomposition temperature, fine particles with better properties of inorganic materials are obtained. If the ligands used to get cation solid solution contain large combustible group, then, large no. of gases evolved during decomposition so that the final product of inorganic materials exhibits better properties than the same materials synthesized by conventional solid state technique. An example of ligand for the synthesis of cation solid solution is hydrazine carboxylate, $N_3H_3COO^-$. It is a fuel rich ligand and hence, it takes atmospheric oxygen during decomposition to get metal oxides.

Experimental Procedure:

Precursor method of synthesis of inorganic materials is explained using Eu^{3+} doped yttrium hydrazine carboxylate compound. Eu^{3+} and Y^{3+} form compounds with hydrazine carboxylate ligand ($N_2H_3COON_2H_5$). These compounds are $Y(N_2H_3COO)_3 \cdot 3H_2O$ and $Eu(N_2H_3COO)_3 \cdot 3H_2O$. Powder XRD patterns of these compounds are very similar to each other (Fig.34) and hence, these compounds form solid solution when $EuCl_3$ and YCl_3 are treated with $N_2H_3COON_2H_5$ at room temperature.

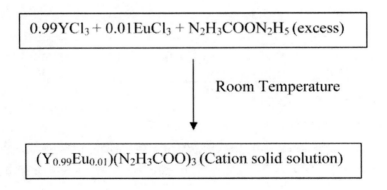

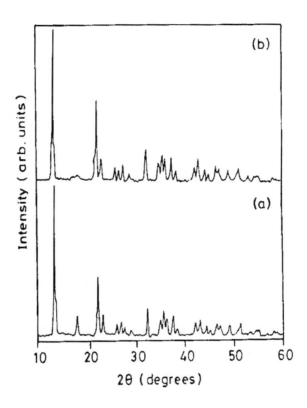

Fig.34: Powder XRD patterns of Y (a) and Eu (b) hydrazine carboxylates

This cation solid solution is decomposed in air in the temperature range 300-1000°C to get red phosphor. Also, presence of oxidizer such as NH_4NO_3 or NH_4ClO_4 reduces the decomposition temperature of the cation solid solution to 300°C.

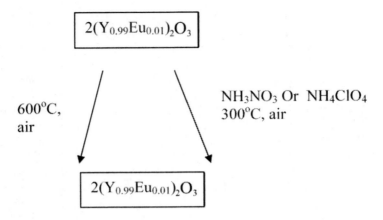

Co-precipitation Method:

General Concept:

This method involves simultaneously precipitating two or more metal salts to form intimate mixtures or solid solutions (precursor). Since water is used as solvent medium it is also called a wet-chemical method. The precursor is obtained either by adding a precipitating agent or by evaporating the solvent. The following equation represents the principle of co-precipitation as intimate mixture of two metal hydroxides.

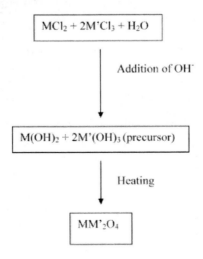

Since precursor is an intimate mixture of two metal ions, formation temperature of MM'$_2$O$_4$ is lowered when compare to that of solid state reaction between M(OH)$_2$ and M'(OH)$_3$. Alkaline hydroxides can be used to get hydroxide precipitate. Other precipitating agents used to get mixed precursors are carbonates and oxalates. My idea in this topic is to explore diamines and hydrazides are basic components to precipitate metal ions as hydroxides.

Experimental Procedure:

Formation of ZnFe$_2$O$_4$ is explained in this section. Zn-oxalate and Fe-oxalate are dissolved in water. Then, water is evaporated to get precursor of oxalates of zinc and iron. Here, the precursor is a solid solution of zinc and iron oxalates. Then, this solid solution is heated to get powder of ZnFe$_2$O$_4$ compound. The following equation represents the formation of ZnFe$_2$O$_4$ powder.

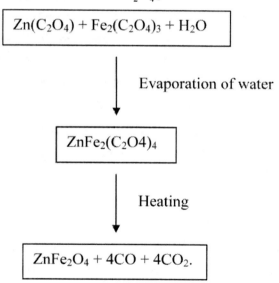

The byproducts are gases and obtained solid product by this method is only ZnFe$_2$O$_4$. A disadvantage of this method is that precipitation of two or more cations, simultaneously, is critical and requires extreme care.

Microwave Synthesis and Sintering of Inorganic Materials:

General Concept:

Solid state synthesis of oxides is a slow reaction by conventional furnace heating method. Therefore, doping of trace quantity of transition elements or rare earth ions in oxide matrices to get technologically useful materials such as phosphors, high temperature superconductors, giant magnetoresitivity materials is limited by conventional solid state reaction. However, the same solid state reaction can be accelerated by starting with reactive precursors such as nano reactants or by flux method or by nonconventional exothermic reactions.

Microwave radiation can also accelerate certain inorganic and solid state reactions. This is due to inherent property of precursors that absorb microwave radiation and convert it into heat energy. Microwave radiation is an electromagnetic radiation and hence, it has both the electrical and magnetic forces that propagate in space with perpendicular direction to each other. Absorption of microwave by certain materials is due to interaction of electrical force of microwave radiation and dipole moment of charged inorganic particles. Absorbed microwave energy by the materials is converted into heat energy by dielectrical loss of materials and hence, the microwave heating of materials is termed as dielectric heating.

Experimental Procedure:

Microwave assisted solid state synthesis of blue lamp phosphor, $BaMgAl_{10}O_{17}:Eu^{2+}$ (BAM) is described as a representative in this section. The detailed description of this can be found in the US patent 7148456.

BAM is conventionally synthesized at 1600-1650°C even in presence of BaF_2 as a flux in a reducing atmosphere. The total duration required to get BAM by the conventional solid state synthesis is more than five hours. The same compound can be synthesized at about 1400°C in about 20 min. Lowering in temperature and duration is the beauty of microwave heating. Reducing atmosphere is required to keep europium ion in +2 oxidation state. The starting materials used for microwave synthesis of BAM are aluminum hydroxide, magnesium oxide, barium carbonate and europium oxide.

As in the short-time synthesis of inorganic materials by microwave heating, microwave can be used to sinter inorganic materials. Thus, microwave sintering leads to shorten the duration. This short-time sintering of porous hydroxyl apatite is explained here to understand the importance of microwave heating in materials chemistry. Well-sintered body with fine grained microstructure exhibits good mechanical property, for example, in the development of good mechanical property of hydroxyl apatite. Conventional sintering of hydroxyl apatite leads to grow the particles and hence, grain sizes of micron are obtained in conventional sintering. Microwave sintering of the hydroxyl apatite yields grain sizes of nanometers. Therefore, better mechanical properties of hydroxyl apatite are obtained in the microwave sintering.

Sol-gel Method:

General Concept:

The sol-gel method is used extensively for the synthesis of silicates and some oxides. It involves three steps and these are

1. Formation of sol
2. Conversion of sol into gel
3. Decomposition of gel

The first step is obtained when molecules combine to yield small particles. Therefore, sol is the dispersion of small particles of oligomers in solvent. The second step is the evaporation of solvent to get extended polymeric gel. This step is crucial to obtain mixed oxides/silicates. The final step is the calcination of gel to get corresponding inorganic materials. This is a method that is free from melt to materials synthesis. This method can be divided into three topics based upon reaction medium and final products.

1. Aqueous sol-gel bulk materials synthesis
2. Nonaqueous sol-gel nanomaterials synthesis
3. Templated sol-gel porous materials synthesis

These topics are discussed below one by one.

1. **Aqueous sol-gel method for bulk materials synthesis:**

    This technique is extensively explored for silicates and other oxides. In general, there are three major steps involved in the synthesis of bulk materials and these are explained using TEOS and water as starting materials.

    First step (Nucleophilic substitution):

    The first step is the nucleophilic substitution of ethoxide groups by water molecules at silicon site. Initially, water molecule attacks silicon site wherein silicon is electron deficient atom. Then, ethanol leaves from silicon site to form OH group attaching with silicon atom. Similarly, other three ethoxide groups leave silicon atom and OH groups substitute for ethoxide groups.

Second step (Condensation reaction):

In the second step, hydrated silicons in tetrahedral environment interact in a condensation reaction forming Si-O-Si bonds. During condensation, water molecule is eliminated and the water molecule occupies in the pore.

Third step (Polycondensation):

In this step, polycondensaiton leads to formation of extended structures and water and ethanol formed in this step occupy in the pores of the network. This extended structure is submicron particles and is termed as sols. Thus, the sols are polymeric networks and are dispersed in water medium. During aging and evaporation of water and ethanol it leads to formation of gel. These steps are explained below (Fig.35)

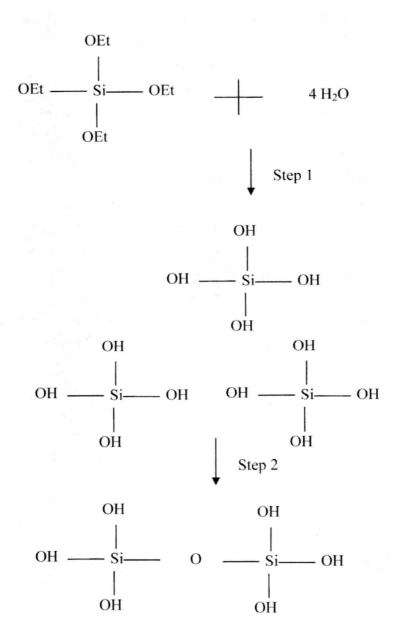

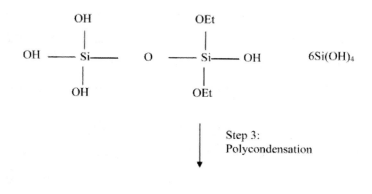

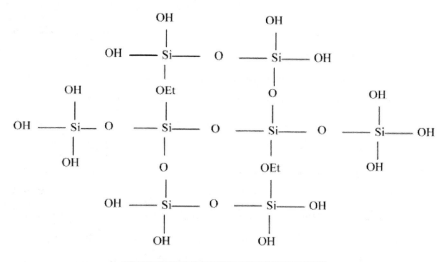

Fig.35: Formation of Si-O-Si bonds

Limitations of aqueous sol-gel method:

Major limitations of aqueous sol-gel method to synthesize nanocrystalline materials are

1. The complex reactions due to high reactivity of metal oxide precursors
2. The double role of water as ligand and solvent
3. In many cases, the three steps such as nucleophilic substitution, condensation and polycondensation occur simultaneously and hence, it is not easy to control them.
4. Slight changes in experimental conditions lead to different particle morphologies.

2. **Nonaqueous sol-gel nano materials synthesis:**

The limitations of aqueous sol gel process can be overcome if water is replaced by organic solvents and hence, nano crystalline materials are possible by non-aqueous sol-gel process. In non-aqueous sol-gel synthesis of metal oxides various kind of reactions leads to M-O-M bond formation. Thus, condensation between metal halides and metal alkoxides forms M-O-M bond by an alkyl halide removal. Similarly, M-O-M bond formation takes place by condensation of two metal alkoxides under elimination of an organic ester. Ester elimination from condensation of metal carboxylates and metal alkoxides gives M-O-M bond formation.

3. **Templated sol-gel porous materials synthesis:**

Various rare-earth oxides with nano tube morphologies have been prepared by template assisted sol-gel method. Because of improvement in properties of nanostructures of rare earth oxides, it is necessary to synthesize them in nano meter sizes.

Research needed:

My idea in this topic is to precipitate rare-earth ions as carbonates by urea hydrolysis in presence of templates (organic compounds) in aqueous medium at relatively low temperature (less than 100°C). This reaction leads to formation of porous carbonates of rare earth ions. After centrifuge this precipitate, careful conversion of carbonates at above 500°C into oxides leads to formation of porous materials. During conversion of carbonates into oxides of rare earth ions organic templates are expelled out.

$NaClO_3$ decomposition after an ignition leads to formation of molten NaCl and gaseous oxygen products. This method of oxygen gas production is extensively explored as emergency oxygen source in aircraft. When the $NaClO_3$ is mixed with metal powder, metal oxides are expected during the exothermic reaction. Therefore, this method of producing oxides is a novel method and production of in situ oxygen might be a fresh oxygen to react with metal powder and to get metal oxides. After the reaction is over, the side product NaCl can be washed out with water and the only solid product is metal oxide.

Electrochemical Methods:

Electrochemical methods are suitable to get coatings on conductive substrate and powder synthesis at near room temperature. There are two distinct electrochemical methods available now to ceramic processing and these are electrophoretic coating and electrolytic coating and powder synthesis. These two methods are described below one by one after a brief description of electrochemical cell construction.

Construction of electrochemical cell:

Mostly, inert and electrical conductive electrodes such as metal/alloy/Pt/graphite are explored as anode and cathode and charged electrolyte. Fig. 36 shows general construction of electrochemical cell.

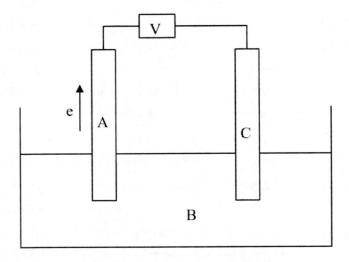

Where A = Anode, V = Power Supply
C = Cathode and B = Electrolyte

Fig.36: Representation of Electrochemical Cell

When a certain potential is applied to the cell, charged particles migrate to electrodes in electrophoretic coating and redox reactions take place in electrolytic powder synthesis. These two methods of electroporetic and electrolytic are described below.

Electrophoretic Coating:

Innert ceramics such as alumina can be coated on substrate by this method. Thus, alumina powder with particle size of less than 30 micron is dispersed in water and ammonium salt of poly(acrylic acid) can be added to water as a charge dispersant. The mixture of alumina powder, ammonium poly(acrylic acid) and water form electrolyte for electrophoretic cell. Commonly employed electrodes are metals on which ceramic coating has been achieved. At this stage, alumina particles become charged by water or charge dispersant and this precursor is ready to migrate by an external electric field. On applying voltage, charged particles migrate to the opposite charged electrodes and form coating on the electrodes. After a certain period of time, the coated electrodes are annealed, if necessary, to crystallize the coated ceramic powder.

This method of electrophoretic coating is cheaper when compare to that of chemimical vapor deposition, physical vapor deposition, plasma sputtering etc.

Electrolytic Method:

Electrolytic method can be explored for coating and powder synthesis. These are achieved by electrogeneration of base in aqueous solution at cathode and oxidation of water soluble metal ions into precipitation of higher valence of metal ions. On applying external voltage, either proton is consumed or hydroxide ion is generated and this chemical reaction is responsible for precipitation of metal ions as hydroxides. Calcination of hydroxides leads to formation of oxides.

Research Needed:

Research in this topic should be directed into microporous or zeolites synthesis and zeolite coating on stainless steel as a corrosion resistant. Also, organically templated open framework materials might be possible at room temperature to synthesize by the electrochemical methods.

CHARACTERIZATION TECHNIQUES

Crystalline Solid:

Crystalline solid has short (atomic level) and long range orders and it is quite different from amorphous solid. Crystalline solid is defined by crystal with definite unit cell. Unit cell is defined as three dimensional group of lattice points that generate whole lattice by translation. Representation of an unit cell is shown below in Fig. 37.

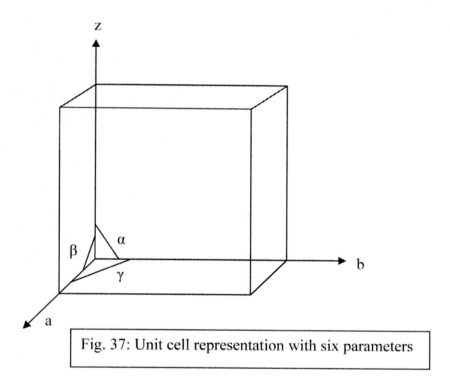

Fig. 37: Unit cell representation with six parameters

Usually, an unit cell is represented by six parameters such as unit cell axes (a, b and c) and three angles between two axes (α, β and γ). There are seven crystal structures possible and these crystal systems with six lattice parameters are tabulated in table 3.

Table 3: Summary of crystal systems and unit cell parameters

Crystal System	Lattice Parameters
Cubic	a=b=c; α=β=γ=90
Hexagonal	a=b#c; α=γ=90, β=120
Tetragonal	a=b#c; α=β=γ=90
Trigonal/ Rhombohedral	a=b=c; α#β#γ#90
Orthorhombic	a#b#c; α=β=γ=90
Monoclinic	a#b#c; α=γ=90#β
Triclinic	a#b#c; α#β#γ

Powder X-ray diffraction technique is valuable technique for solid state and inorganic materials scientists. Even though, it has long history, now-a-days it is a main and fast technique for the characterization of mainly inorganic materials. Before going to describe the importance of powder XRD let me very briefly explain the derivation of Bragg's equation. This equation is simple but it explains depth of structure of polycrystalline and single crystal materials. All scientists involving in solid state and inorganic materials synthesis rely on this equation.

Bragg's Equation:

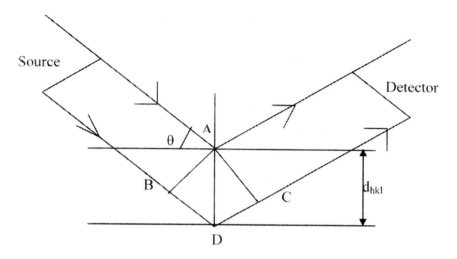

Fig.38:. Diffraction of X-rays by a crystalline material

Wavelength of X-rays is in the order of difference between lattice planes of crystals. Therefore, crystals behave as grating for X-rays. Hence, X-rays are used to solve structure of crystals.

Let us consider that atoms or ions as points arranged in a regular fashion in planes. When the polycrystalline powders are exposed to X-ray with an unique wavelength, λ it is diffracted by array of ions (Fig.38). If diffracted waves are in-phase, amplitude of them is doubled from the incident wavelength. If diffracted waves are out of phase, then amplitude of them is zero. Therefore, detector sees diffracted waves only if they are in-phase.

Let us assume the angle of incident wavelength (A and D points) of X-ray at the plane of atoms in a crystal is θ.

The path difference between two lattice planes depends upon wavelength of X-ray and it is equal to $n\lambda$ \hfill (1)

BD and DC are the extra distances of X-ray traveled for the diffraction at D from diffraction at A. Therefore, path difference = BD + DC = $2d_{hkl} \sin\theta$ \hfill (2)

Comparing equations (1) and (2), we get
$$n\lambda = 2d_{hkl} \sin\theta \quad (3)$$
This equation is called Bragg's equation.
Rewriting the equation (3),
$$\lambda = 2d_{hkl} \sin\theta/n \quad (4)$$
n = 1 for a crystalline system. Then, equation (4) becomes
$$\lambda = 2d_{hkl} \sin\theta \quad (5)$$

Note: Reason for n = 1

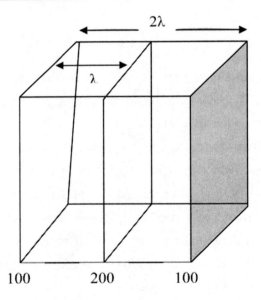

Fig.39: Diagram of 100 and 200 planes in an unit cell

The Bragg equation is $d_{hkl} = n\lambda/2\sin\theta$ \hfill (6)

For second order diffraction from 100 plane, the Bragg equation becomes

$$d_{100} = 2\lambda/2\sin\theta = \lambda/\sin\theta \tag{7}$$

For the first order diffraction from 200 plane, the Bragg equation becomes

$$d_{200} = \lambda/2\sin\theta \text{ or}$$

$$d_{100}/2 = \lambda/2\sin\theta = \lambda/\sin\theta \tag{8}$$

Equation (7) = Equation (8).

X-ray diffraction:

The powder X-ray diffraction patterns of solid materials are obtained using X-ray diffractometers using Ni filtered Cu- Kα radiation. The X-ray generator is operated usually at 30 kV and 30 mA. High purity Si powder is used as the internal standard.

Determination of average crystallite size:

The average crystallite size (D) from X-ray line broadening is calculated using the Scherrer equation.

$$D = 0.9\lambda/\beta_{1/2}\cos\theta$$

Where λ is the wave length of the X-ray beam, $\beta_{1/2}$ is the angular width at the half maximum intensity and θ is the Bragg angle. The instrumental broadening is corrected using quartz as internal standard

Thermal Analysis:

Thermal analysis (TA) encompasses a group of techniques in which physical and or chemical properties of substance are measured as a function of temperature whilst the substance is subjected to a controlled temperature change (heating or cooling). These include thermogravimetry (TG), differential thermal analysis

(DTA), differential scanning calorimetry (DSC), thermo-mechanical analysis (TMA) etc. The techniques, parameter measured and the instrument used are listed in table 4.

Table 4: Methods of thermal analysis

Technique	Parameter measured	Instrument
Thermogravimetry	Weight change	Thermobalance
Derivative Thermogravimetry	Rate of weight change	Thermobalance
Differential thermal analysis	Temperature difference between the sample and the reference	DTA apparatus
Evolved gas analysis	Amount of gas liberated	GAS analyzer

In many cases, use of a single technique may not provide sufficient information and hence, the use of other thermal analysis techniques either independently or simultaneous measurements for complementary information becomes necessary.

Simultaneous TG-DTA:

Thermogravimetry (TG) is a technique in which a change in weight of a substance is recorded as a function of temperature or time. In a typical thermogram the following features could be noted.
1. The horizontal portions (plateaus) indicate the regions where there is no weight change.
2. The curved portions are indicative of weight loss.
3. Since the TG is quantitative, calculations on compound stoichiometry can be made at any given temperature.

In differential thermal analysis (DTA) both the test sample and an inert reference material (usually alumina) undergo a controlled heating or cooling program which is usually linear with respect to time. There is zero temperature difference between the sample and the reference material when the former does not undergo any physical or chemical change. If, however, any change (endothermic or exothermic) takes place, a difference in temperature (ΔT) will occur between the sample and the reference material. In DTA, a plot of ΔT vs temperature or time is made. The shape and size of the peaks in a DTA curve can give information about the nature of the test sample. Thus, sharp exothermic peaks often signify changes in crystallinity or fusion processes, whereas broad endotherms arise from dehydration reactions. Physical changes usually result in endothermic peaks whilst chemical reactions, particularly those of oxidative nature, are predominantly exothermic.

Scanning electron microscopy:

The size and distribution of the grains of the sintered samples are examined using scanning electron microscopy. The SEM is usually explored for measuring grain sizes and it is operated at 20 kV. The samples were made conducting by sputtering them with Au-Pd alloy. The chemical composition of the sample is estimated by energy dispersive X-ray spectroscopy (EDXS).

Transmission electron microscopy:

Transmission electron microscopic studies are performed to ascertain the particle morphology and the extent of agglomeration. Transmission electron microscopy is usually operated at 200 kV. Specimens are prepared by suspending the sample in acetone or ethanol and allowing a drop of the resulting suspension to dry on a copper grid. The particle size of materials is determined by averaging out the data obtained in a series of batches.

Optical spectroscopy:

Infrared (IR) spectroscopy has been used to identify group frequencies in general and the M-O stretching frequencies in particular of oxides. Infrared spectra of solid samples in the region

$300 - 1400$ cm^{-1} are recorded. In the case of organically templated open framework materials, infrared spectroscopy is very important to identify the organic moiety in the inorganic compounds.

When light strikes a phosphor in the ground state, it absorbs radiation of certain specific wavelength to jump to an excited state. A part of the excitation (absorbed) energy is lost on vibrational relaxation, i.e., radiationless transition to the lowest vibrational level takes place in the excited state. And eventually it returns to the ground state by emitting energy, which is called fluorescence. Fluorescence continues for a period of 10^{-8} to 10^{-9} sec in most cases. Since a part of the radiation absorbed is lost, the fluorescence emitted from the substance has a longer wavelength (lower energy) than the excitation radiation (Stokes' law). The emission transition occurring in a solid is seen as a glow and is registered in the form of a band in the luminescence spectrum. The position of the band in the luminescence spectrum does not depend upon the method of excitation. The luminescence spectra are normally observed with the intensity of luminescence as a function of the emission wavelength. The same instrument can be used to measure the spectral distribution of luminescence (emission spectrrum) and the variation in the emission intensity with excitation wavelength (excitation spectrum) or with activator concentration.

STRUCTURE OF INORGANIC MATERIALS

Bonding of atoms to form a compound plays important role in determining the structure, physical and chemical properties of them. For example, both the graphite and diamond are made up of carbon. But, their properties are completely different from each other. Thus, graphite is a soft, slipper material used as a lubricant in locks and diamond is one of the hardest material known now, which finds applications as gemstone and industrial cutting tools. Similarly, Si and C are the same group, but their properties are completely different from each other when they combine with oxygen. Thus, CO_2 is a gaseous molecule and SiO_2 is a solid with a extended structure.

Types of chemical bondings:

There are three types of chemical bonding known now and these are ionic, covalent and coordination bondings. All the bondings occur due to the interaction of outermost electrons of atoms. Hydrogen bonding is also known but, it is not included here in this section. For the coordination bondings please refer to coordination complexes mentioned in the earlier chapter.

Ionic bonding:

Ionic bonding occurs between ions. The typical example is NaCl. Sodium atom donates its outermost electron to neighboring chlorine to form Na^+ ion. Chlorine atom attains the extra electron that comes from sodium to form Cl^- ion. Na^+ and Cl^- ions attain noble gas electronic configuration. Thus, Na^+ and Cl^- ions have mutual electrostatic interaction for the formation of ionic bonds. This is the driving force for the ionic bonds. Ionic bonding occurs between metals and non metals.

Structures of AB compounds:

Structure of NaCl:

The structure of NaCl is a simple cubic with Cl⁻ ion occupies at the corner and face centered of a simple cube whereas Na⁺ ion occupies at the edges of cubic system. The coordination number is six for both the Na⁺ and Cl⁻ ions. The ionic radius ratio, r_{cat}/r_{anion} fot this structure is between 0.414 and 0.732. For NaCl, the ratio of radii is 0.54. Some ionic compounds belonging to NaCl structure are LiCl, KCl, RbCl, NH_4I and alkaline earth metal oxides. The structure of NaCl is represented below in Fig. 40.

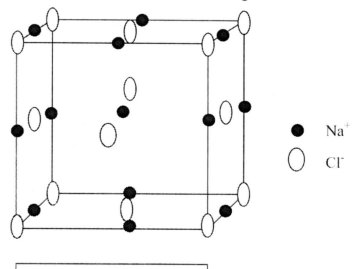

Fig.40: Structure of NaCl

Structure of CsCl:

This is also cubic structure and Cl⁻ ion occupies each corner and the large ion, Cs⁺ is fixed at the centre of cube. Thus, coordination number of Cs⁺ ion is eight and for Cl⁻ ion it is six. Because of Cs⁺ ion is bigger ion than Na⁺ ion, Cs⁺ ion does not occupy at the edge of the cube. Radius ratio of CsCl structure is $1 > r_+/r_- > 0.732$. For CsCl, it is 0.93. Transition metal ions are smaller in size and therefore, they do not form CsCl structure. Other compounds having CsCl structure are CsBr, CsI, NH$_4$Cl and NH$_4$Br. The CsCl structure is given below in Fig. 41.

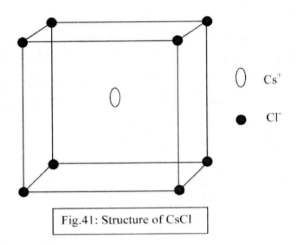

Fig.41: Structure of CsCl

Structure of Zinc blende:

The radius ratio of ZnS is 0.40. Therefore, each Zn^{2+} cation is surrounded by four sulphide anions in a tetrahedral fashion. Similarly, each sufide anion is surrounded by four zinc cations in tetrahedral geometry. The following Fig.42 shows zinc blende structure.

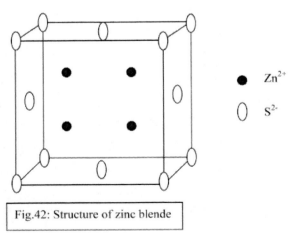

Fig.42: Structure of zinc blende

It can be viewed that sulfide anion occupies corners of cube and in certain faces whereas zinc cation occupies at the centre of divided each cube.

CdSe naocrystal with zinc blende structure is useful in hybrid solar cells. This compound is prepared by oleic acid modified wet-chemical method. Spherical inorganic nanocrystals are obtained by this method. When it is as an acceptor combined with polymer, poly(2-methoxy-5-(2'-ethylhexoxy)-P-phenylenevinylene as a donor shows a good energy conversion efficiency in hybrid solar cell.

Structure of Wurtzite:

Wurtzite structure is another form of ZnS structure named after French Chemist Charles-Adolphe Wurtz. Unlike zinc blende structure, wurtzite is a member of the hexagonal crystal system and it consists of tetrahedral coordinated zinc and sulfur ions. These atoms are stacked in an ABABAB pattern. Compounds that fall in this category are AgI, ZnO, CdS and so on.

ZnO doped with cobalt shows visible light driven oxygen evolution from aqueous solution containing silver nitrate as electron scavenger. The cobalt doping in ZnO can be achieved by solution combustion method using 3-methyl-pyrazole-5-one as a fuel at 350°C. The doping level in zinc oxide can be as high as 10 mol.% without affecting the crystal structure.

CdS semiconductor can be incorporated in the interlayer of hecotrite and this nanocomposite of CdS/hectorite exhibits good catalytic activity for hydrogen production from aqueous solution containing methanol as hole scavenger at room temperature. The formation of nanocomposite is usually confirmed by powder XRD pattern and DRS.

Structures of AB$_2$ type compounds:

Compounds having AB$_2$ type composition have two different structures and these are fluorite and rutile structures. These structures are described below.

Structure of Fluorite:

In this structure, cation has eight coordination environment whereas anion has tetrahedral coordination. Thus, anion occupies at the corners of the cube whereas cation is fixed at the alternate centre of the cube or half the body centre position (Fig. 43). If the positions of the cations and anions are reversed, antifluorite structure is obtained.

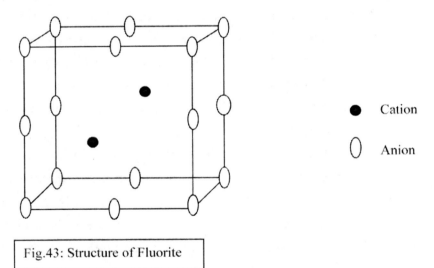

Fig.43: Structure of Fluorite

Structue of Rutile:

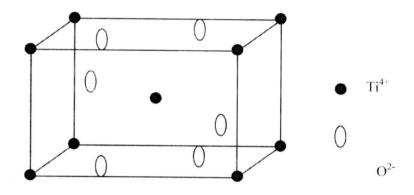

Fig. 44: Structure of rutile

The rutile structure in which most of the transition metal dioxides crystallize is shown in the above Fig. 44. Thus, transition metal ion occupies at the corners of the cube and centre of the cube whereas oxide anion is fixed at some faces of the cube as shown in the above figure. The positions of cation and anion in the rutile structure lead to six coordination for Ti^{4+} cation and three coordination for oxide anion which is positioned corner of triangle.

Anatase titanium dioxide is a low temperature phase of TiO_2 compound. It crystallizes as tetragonal structure and the higher photocatalystic activity of anatase over rutile TiO_2 is due to higher surface area. Therefore, there has been extensive research activity on anatase titania and its uses as photosplitting of water.

Structure of Perovskite:

Perovskite structure was named after the discovery of $CaTiO_3$ by the Perovski scientist. This structure is represented by ABO_3 nominal composition. In this structure, the size of A cation is larger than that of B-cation. Variety of compounds exhibiting this structure finds plethora applications, for example high temperature superconductors.

The coordination geomentry of B cation is octahedral whereas that of A cation is 12. The perovskite structure is shown in Fig. 45.

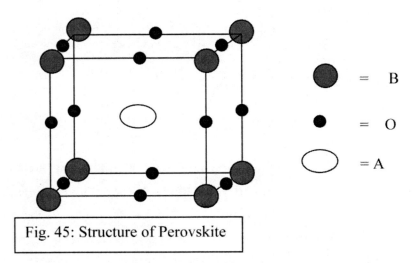

Fig. 45: Structure of Perovskite

Thus, A cation occupies at the centre of cube and B cation occupies at the corner of the cube where as oxide anion is fixed at the edge of the cube as shown in the figure above. Examples for different perovskite compounds are $LaAlO_3$, $SrSiO_3$ and $NaWO_3$.

One important compound falling perovskite structure is $YBa_2Cu_3O_{6.9-7.0}$. This is an important compound since it has property of zero resistance in electrical conductor at liquid nitrogen temperature. Therefore, there has been extensive study on this High Tc Superconductor. It is necessary to post treatment this HTc after it is formed to get the novel property of HTc. This HTc structure is derived from three pervoskite structures and the final structure obtained by this way is shown Fig. 46.

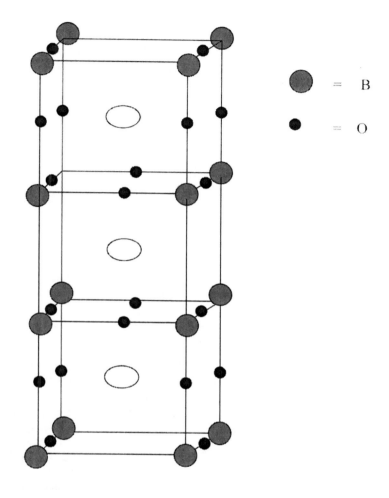

Fig. 46: Structure of HTc superconductor, $YBa_2Cu_3O_7$

Perovskite phosphor:

Yet another important feature of this structure is Eu^{3+} doped $GdAlO_3$ perovskite phosphor. This red emitter could be synthesized by solution combustion process at 500°C using diformyl hydrazine as a fuel. Emission spectrum of Eu^{3+} doped $GdAlO_3$ is compared with that of Y_2O_3:Eu,Zn red phosphor in the following Fig. 47.

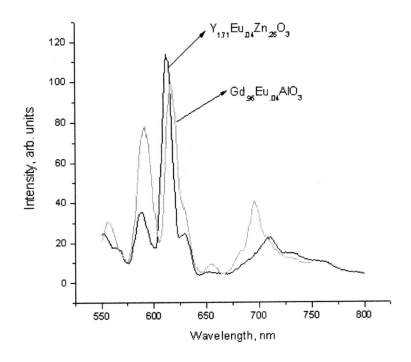

Fig. 47: Emission spectra of Eu^{3+} in two different hosts

There are two strong emission peaks observed for $GdAlO_3:Eu^{3+}$. One is at 615 nm and another emission is at 592 nm. The perovskite red emitter has equal intensity with that of Y_2O_3:Eu,Zn red emitter. This is the first perovskite red emitter exhibiting both the transition synthesized by solution combustion method. Perovskite red emitter is a special phosphor because it shows simultaneous emission peaks due to $^5D_0 \rightarrow {}^7F_2$ (615 nm) and $^5D_0 \rightarrow {}^7F_1$ (592 nm) transitions. Because of this special property of perovskite red emitter, it has both higher CRI and greater lumen output.

Structure of Spinel:

Compound of spinel structure with nominal composition is represented as AB_2O_4. Thus, to balance oxide anion charge as 8-, there are 3 ways by which two different cations combine to give 8+. Some examples are $NiFe_2O_4$, $TiMg_2O_4$ and Na_2WO_4. The structure of normal spinel is shown in Fig. 48.

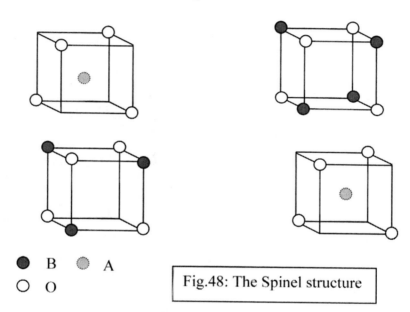

● B   ◎ A
○ O

Fig.48: The Spinel structure

The unit cell in figure 48 shows the tetrahedral and octahedral co-ordination of the A and B cations. Zn_2TiO_4 belongs to normal spinel. Therefore, coordination number of Zn ion is 4 and that of Ti ion is 6. In the case of inverse spinel, the formula is $B^t(AB)^oX_4$. Zn_2SnO_4 belongs to inverse spinel. Smallar cation occupies tetrahedral site and larger cation occupies octahedral site.

Special structures of inorganic oxides:

Y_2O_3:

Yttria, Y_2O_3 crystal structure is shown in Fig. 49. It belongs to cubic bixbyite type structure. The cubic bixbyite lattice is derived from the fluorite lattice with the removal of 2 oxygens either along the face or body diagonals leading to the two crystallographically inquivalent sites with S_6 and C_2 symmetires for Y^{3+} ion. The relative concentration of these two (S_6 and C_2) sites is in a 1:3 ratio respectively. This structure is stable at all temperature up to 2310°C. Most of rare earth oxides have structures similar to Y_2O_3.

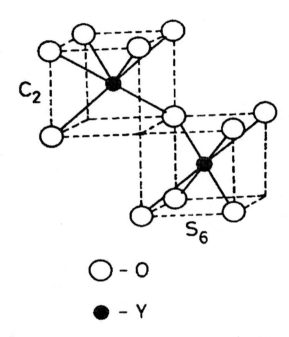

O - O

● - Y

Fig. 49: Structure of Y2O3.

This compound has been explored as a host for Eu^{3+} rare-earth ion. Eu^{3+} ion has equal probability for occupying C_2 and S_6 sites. Presence of Eu^{3+} in yttrium site is usually confirmed by fluorescence spectrum. The emission spectrum of the Y_2O_3:Eu^{3+} shows the characteristic emission at 611 nm (Fig. 50). That is, Eu^{3+} ions have occupied Y^{3+} ions at C_2 site in the Y_2O_3 lattice. This characteristic emission band is attributed to the electric dipole transition, $^5D_0 \rightarrow ^7F_2$ of the Eu^{3+} ion. Emission due to Eu^{3+} ion from S_6 site is also seen in the emission spectrum. But, the intensity of 611 nm peak is higher than that of the other bands. This is usually observed if the product is homogeneous which favor strong energy transfer from S_6 to C_2.

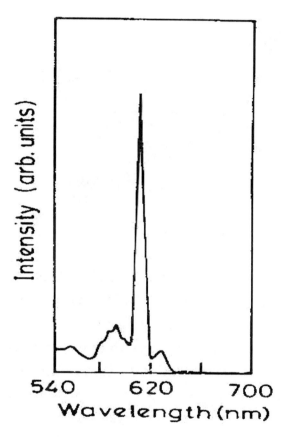

Fig. 50: Emission spectrum of Eu^{3+} ion in Y_2O_3.

Structure and luminescence of YAG doped with Ce^{3+}:

$Y_3Al_5O_{12}$ possess garnet structure. In the garnet structure, yttrium ion occupies 8-coordination, aluminum ions have six and four coordination environment. Ce^{3+} occupies at Y^{3+} site and the coordination number, therefore, for Ce^{3+} ion is eight. Fig. 51 shows crystal structure of YAG.

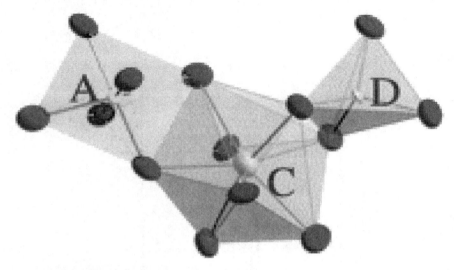

Fig. 51: Structure of YAG, where A&D = Al, C = Y

There are two Ce^{3+} absorption bands in the region between 250 and 500 nm observed. They center at 342 and 462 nm due to the crystal field splitting of the 5d orbitals. Obviously, the broad band from 400 to 500 nm is the most intense, which makes it possible for the phosphor at apply with blue InGaN LED. But, there is only one broad emission band located from 450 to 650 nm, which is an ideal yellow light that complements the blue light emitted by InGaN chip to generate white light. Fig. 52 shows the emission spectrum of Ce:YAG under the blue LED.

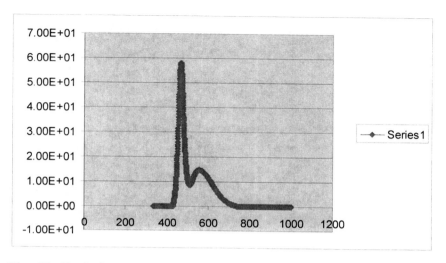

Fig. 52: Emission spectrum of YAG:Ce under blue LED

Drawback of Ce:YAG:

Even though Ce:YAG is a promising material, it requires as high as 1300°C for the solid state synthesis. Because of high temperature synthesis cost of the material becomes high.

Research needed:

Therefore, it is highly desirable to reduce the synthesis temperature of Ce^{3+} doped phosphors. This could be achieved by replacing refractory Al_2O_3 by transition metal oxides. Therefore, I would like to draw your attention toward niobates and tantalates, having garnet structure.

Proposed Oxides with garnet structure:

(1) Li_2O-AO-La_2O_3-Ta_2O_5, where A = Sr, Ba.
(2) Li_2O-AO-La_2O_3-Nb_2O_5, where A = Ca, Sr and Ba
(3) Li_2O-La_2O_3-M_2O_5, where M = Nb and Ta.

Solid state synthesis of the third series of garnet phosphors:

$Li_2CO_3 + La_{2-0.05}O_3 + 0.05CeF_3 + Nb_2O_5/Ta_2O_5 + 10$ wt.% excess Li_2CO_3

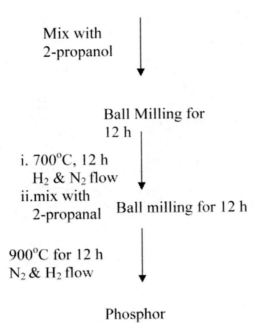

Other Possibilities:

1. In the first and second series of proposed oxides, Eu^{2+} might be successfully doped at the 8-coordinated alkaline earth ions to get LED phosphors.

2. $CaO-Al_2O_3-SiO_2$ was the first mineral identified as a garnet structure. In this structure the 8-coordinated Ca^{2+} might be replaced by Eu^{2+} ion to get LED phosphor.

3. Tetrahedral silicon can be substituted in $CaO-Al_2O_3-SiO_2$ by boron and for charge compensation Ca is replaced by yttrium. The resultant garnet compound is $Y_2O_3-Al_2O_3-B_2O_3$. Solid state synthesis of this compound is reasonable lower than that of $Y_3Al_5O_{12}$ compound. And also, Ce^{3+} can be doped at Y^{3+} in $Y_2O_3-Al_2O_3-B_2O_3$ to get blue LED excitable phosphor.

IMPORTANT TOPICS

PHOTOCATALYSTS FOR WATER SPLITTING

Introduction:

There are two vital applications of semiconductors in the chemistry field. One is the conversion of photon energy into electrical energy or chemical energy. Another application is the water purification. Photosplitting of water into H_2 and O_2 is an up-hill reaction and it is a reversible reaction. Therefore, ΔG is greater than zero. Whereas detoxification of organic compounds in aqueous medium is down-hill reaction and it is an irreversible reaction. Therefore, ΔG is less than zero.

Principle of Photosplitting of water using semiconductors:

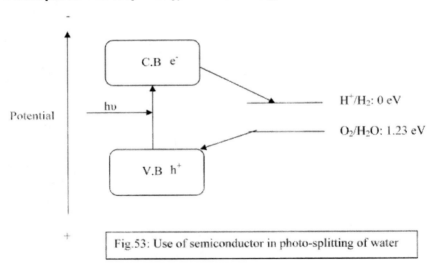

Fig.53: Use of semiconductor in photo-splitting of water

Level of bottom of conduction band should be more negative than that of redox potential of water into hydrogen, and level of upper of valence band should be more positive than that of redox potential of water into oxygen for a semiconductor to be a photosplitting of water into H_2 and O_2 (please refer to Fig. 53). The potential difference between these two levels (of H^+ into H_2 and H_2O into O_2) is 1.23

eV. Therefore, semiconductor with band gap greater than 1.23 eV is required. When a semiconductor is exposed to wavelength greater than that of band gap, charge carriers are produced. Electrons are promoted to the CB whereas holes remain in the VB. These charge carriers are responsible for photosplitting of water into H_2 and O_2 as shown in the Fig.53. Sometime sacrificial reagents are used to get either H_2 or O_2. In a real system, there are a few reactions taking place. Once, charge carriers are formed in semiconductor particles, charge carriers should migrate to the surface of particles (charge separation) before they recombine. Then, the charge carriers involve in the photosplitting of water.

Classical photocatalyst, TiO_2:

TiO_2 is a well studied photocatalyst in water splitting and pollution purification as well. Various studies on this catalyst reveal important conclusions and still research on this catalyst is in progress. This is mainly due to chemists that engineered the classical catalyst into various forms, and to study the effect of various forms on the photocatalysis. Three different crystal structures have also been explored to investigate these structures on photocatalysis. There is lot of experimental results available in the literature to compare, conclude and direct the future of the titania photocatalyst. Therefore, in the following sections, the crystal structures, various syntheses, different forms will be described in detail but in relevant to photosplitting of water using classical photocatalyst, TiO_2.

Crystal Structures of TiO_2:

There are two well known crystal structures for TiO_2, and these are anatase and rutile. Anatase is a low temperature phase where as rutile is a high temperature phase. In the both structures titanium is octahedral coordination. These structures are shown in Figs. 54 and 55. But, low temperature phase, anatase titania with high crystallanity is known to show better catalytic activity than that of rutile phase. This is due to less defects in high crystallanity in anatase. When anatase transforms into rutile in the temperature range 500-600°C, crystal growth takes place which results in more defects in rutile phase. These defects are centre for recombination of charge carriers and hence, rutile TiO_2 has lower photocatalytic activity than that of anatase TiO_2.

Fig.54: Polyhedral representation of anatase TiO2

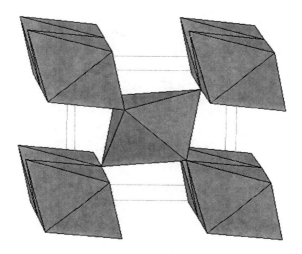

Fig. 55: Polyhedral view of rutile TiO2

Structure of $K_4Nb_6O_{17}$ and other information available in the literature:

$K_4Nb_6O_{17}$ is a layer compound and it is made up of NbO_6 octahedral layer and two kinds of interlayers. One interlayer is readily hydrated whereas another interlayer is hardly hydrated at room temperature. But, both the interlayers are hydrated in aqueous solution. However, in the both interlayers, potassium cation occupies. The polyhedral representation of $K_4Nb_6O_{17}$ is shown in Fig. 56.

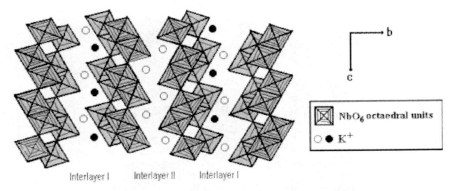

Fig.56: Polyhedral representation of K4Nb6O17

$K_4Nb_6O_{17}$ was a thoroughly studied semiconductor for photosplitting of water. It was characterized by various physical techniques such as FT-IR, EXAFS, TEM, SEM and TPR. It was reported earlier that photosplitting of water was more in aqueous solution of sodium carbonate using $K_4Nb_6O_{17}$/Pt semiconductor. Similar to Pt, Ni loaded $K_4Nb_6O_{17}$ showed good photocatalytic activity for water splitting. Effect of particle sizes of $K_4Nb_6O_{17}$ on photosplitting of water was also studied in detail.

Commercial synthesis of TiO_2:

It has been achieved by converting titanium ore, viz. ilmenite into $TiCl_4$ by the following equation.

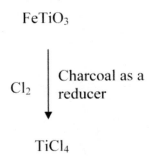

When ilumenite is chlorinated in presence of charcoal reducer, low temperature volatile $TiCl_4$ is obtained. The purpose of charcoal is to reduce iron(II) ion into iron metal. If iron is also chlorinated into $FeCl_3$, the fractional distillation is employed to separate $TiCl_4$ from $FeCl_3$. The conversion of $TiCl_4$ into TiO_2 takes place by the following equation.

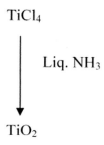

$TiCl_4$ on hydrolysis yields titania. This is how titania is commercially produced.

Chemistry of nanocomposite semiconductors in hydrogen economy:

Nanosemiconductors are useful in production of hydrogen gas from water under visible or UV irradiation. The hydrogen economy in this direction was originally started by Prof. Tsugio Sato in Japan. Chemistry involved in synthesizing nanosemiconductors is very interesting and the steps to synthesize nanosemiconductors are explained in this section. There are two possible ways by which nano titania could be synthesized. One is the incorporation of nano titania in the interlayer of insulator layers, and another is nano titania in the interlayer of semiconductors. This is explained in the flow chart below. Both the composites are proven to be efficient to photosplitting of water

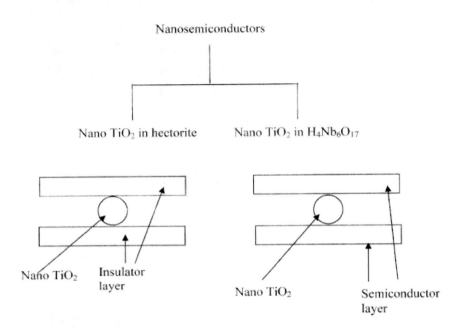

Chemical synthesis of nano TiO$_2$ in hectorite:

Hectorite is an insulator layer compound and sodium cation occupies in the interlayer of hectorite. It is chemically possible to ion-exchange sodium cation by other positive ions such as $[Ti(OH)_x(CH_3COO)_y]^{z+}$ (Titanium acylate). Titanium acylate is a cation stable in acidic medium and therefore, it is usually prepared from titanium isopropoxide precursor in excess of water and acetic acid. It is very interesting observation to mention here that when hectorite is put into distilled water, hectorite dissociates into a clear solution. This is due to hydration of sodium cations present in the interlayer. However, when hectorite is put into acidic solution of titanium acylate, hectorite remains stable. Therefore, it is possible to ion-exchange part of sodium cations present in hectorite by titanium acylate cation molecule. After ion-exchange reaction of hectorite is over, titania in hectorite is usually obtained either by calcinations at high temperature or by photodecomposition of titanium acylate at room temperature. The calcination temperature required to decompose titanium acylate is obtained by simultaneous TGA-DTA experiment (Fig.57).

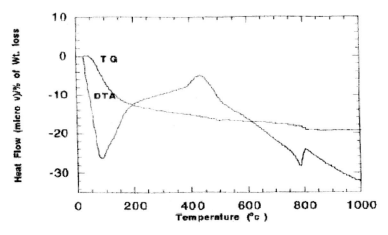

Fig.57:Simultaneous TGA-DTA

The following flow-chart describes steps involved to synthesize nano TiO_2 in hectorite.

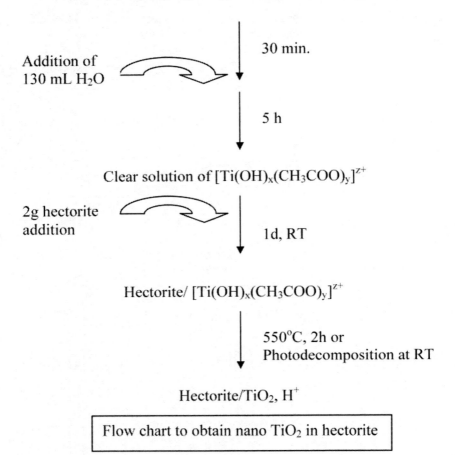

Flow chart to obtain nano TiO_2 in hectorite

The final product, nano TiO_2 in hectorite contains proton for charge compensation. Thus, this compound or similar compounds open up new topic of proton conductors which are stable even at 500°C.

Evidence for formation of nano TiO_2 in hectorite:

Diffuse reflectance spectra of hectorite/Ti-acylate heated at 550°C and 1000°C are shown in Fig. 58.

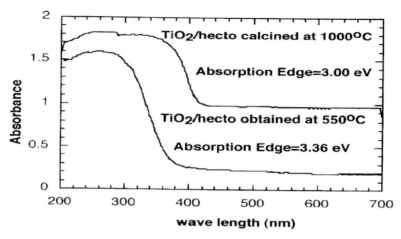

Fig.58: Comparison of DRS of nano and bulk TiO_2 in hectorite

It is clearly understood that the band gap of hectorite/Ti acylate heated at 550°C is greater (3.36 eV) than that of hectorite/Ti acylate heated at 1000°C (3.00 eV). This blue shift observed in DRS clearly evidences the formation of nano TiO_2 and this observation is called quantum size effect. It is also interesting to note that the DRS intensity of nano TiO_2 in hectorite is greater than that of bulk TiO_2 in hectorite.

Photoproduction of hydrogen gas from aqueous methanol solution using nano TiO_2 in hectorite:

Nano TiO_2 in hectorite is capable of hydrogen production from water and the photoactivity of the materials is enhanced by co-incorporation of Pt. Fig. 59 compares photoproduction of hydrogen gas from aqueous methanol solution under UV irradiation using nano TiO_2 in hectorite and nano TiO_2 and Pt in hectorite.

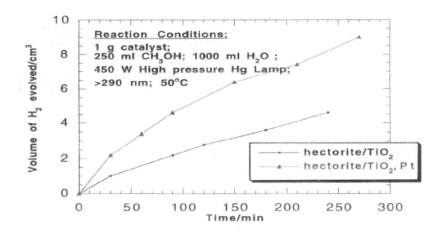

Fig.59: Photoproduction of hydrogen gas using nano TiO2 in hectorite

The important observations are that either hectorite or hectorite/Pt does not show photocatalytic activity for hydrogen production under the same conditions employed for nano TiO_2 in hectorite. The other valuable note is that the photoproduction of hydrogen gas using nano TiO_2 in hectorite is greater than that of commercial P-25 photocatalyst although the TiO_2 contents in hectorite/TiO_2 is only 4.63%.

Chemical synthesis of $K_4Nb_6O_{17}$ based nano semiconductor composites:

Fusion reaction between K_2CO_3 and Nb_2O_5 with molar ration 2:3 respectively, is carried out at 1200°C for 20 min. to get $K_4Nb_6O_{17}\cdot 3H_2O$. Then, $K_4Nb_6O_{17}\cdot 3H_2O$ is refluxed with 5 M HCl for 30 min. which leads to ion-exchange reaction to get $H_4Nb_6O_{17}$. Then, acid-base reaction between $H_4Nb_6O_{17}$ and $[Pt(NH_3)]^{2+}$ results in formation of $H_4Nb_6O_{17}/[Pt(NH_3)_4]$. Then, photodecomposition of $H_4Nb_6O_{17}/[Pt(NH_3)_4]$ leads to formation of $H_4Nb_6O_{17}/Pt$. At this stage, $[Ti(OH)_x(CH_3COO)_4]^{z+}$ can not be ion-exchanged with $H_4Nb_6O_{17}$ due to difficulty for expansion of $[Nb_6O_{17}]^{4-}$ layer. Therefore, reaction between $H_4Nb_6O_{17}$ and $C_3H_7NH_2$ at RT for 6 h results in expansion of interlayer of $H_4Nb_6O_{17}$. The driving force for this reaction is acid-based reaction. Then, Ti-acylate is successfully incorporated in the interlayer of $H_4Nb_6O_{17}$, and Ti-acylate is photochemically decomposed into nano TiO_2 in $H_4Nb_6O_{17}$. The following flow chart illustrates the wet-chemical synthesis of $H_4Nb_6O_{17}/Pt,TiO_2$.

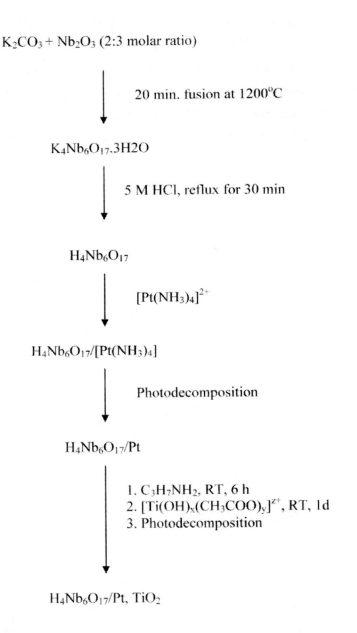

Flow chart for the chemical synthesis of nano TiO_2 in $H_4Nb_6O_{17}$

Evidence for the formation of nano TiO_2 in $H_4Nb_6O_{17}$:

Powder XRD technique is explored to follow the various formation of semiconductor composites. Fig. 60 explains the shift in low angle powder XRD peak. Thus,

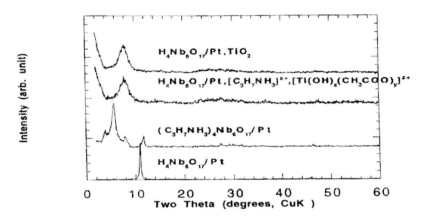

Fig.60: Powder XRD patterns of various semiconductor composites

When big molecule, $C_3H_7NH_2$ is treated with $H_4Nb_6O_{17}$, the low angle peak further shifts into lower angle. This is due to expansion of $[Nb_6O_{17}]^{4-}$.

Photoproduction of hydrogen gas from aqueous methanol solution using nano TiO_2 in $H_4Nb_6O_{17}$:

$K_4Nb_6O_{17}$ shows photoevolution of hydrogen gas similar to P-25. The photocatalytic activity of $H_4Nb_6O_{17}$ is 3.5 times greater than that of $K_4Nb_6O_{17}$. Interestingly, photocatalytic activity of $H_4Nb_6O_{17}$ is improved to 4.2 and 8.5 times by the incorporation of Pt and co-incorporation of TiO_2 and Pt, but it is slightly decreased by incorporation of TiO_2 alone. Fig. 61 compares photocatalytic activity of $K_4Nb_6O_{17}$ based semiconductors.

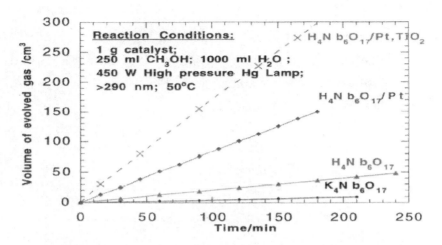

Fig.37: Photoproduction of hydrogen gas using $K_4Nb_6O_{17}$ based semiconductors.

DILUTED MAGNETIC SEMICONDUCTORS

Diluted magnetic semiconductors (DMS), also referred to as semimagnetic semiconductors, currently receive great attention owing to their potential applications in spintronics, magnetic switches and magnetic recordings. To obtain DMS, lattice ions in semiconductors are partly substituted by magnetic elements, such as transition metal or rare-earth (Fig.62).

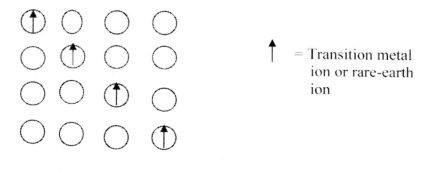

Fig.62: Representation of DMS.

In DMS quantum dots (characteristic size <100 nm), spins are used as information carriers and therefore, in addition to charge, electron spin is controlled to introduce new functionality in semiconductor devices. The most extensively studied DMS are $A^{II}_{1-x}Mn_xB^{VI}$ alloys where A = Cd, Zn, Hg etc. and B = S, Se, Te. Some examples are $Cd_{1-x}Mn_xSe$, $Zn_{1-x}Mn_xS$ and $Hg_{1-x}Mn_xTe$. More recently, attention has been focused primarily on oxides which provide superior properties, including stability. In this context, bulk and nano DMS have been realized in $Zn_{1-x}Mn_xO$, $Zn_{1-x}Co_xO$ and $Zn_{1-x}Ni_xO$. The ferromagnetic properties of DMS stem from the interaction of lattice, spin and electronic degrees of freedom between magnetic ions and semiconducting bulk/nano materials. Prediction of room-temperature ferromagnetism on transition metal ions doped

ZnO by computations was verified by the preparation of thin films of Co^{2+}, Fe^{2+}, Mn^{2+} and V^{2+} doped ZnO.

Despite the recent progress in studies of DMS, some challenges remain for their production. First, it is difficult to reach uniform distribution of dopants in semiconductors for the required small concentrations. Another problem is segregation of transition metal oxides from solid solution due to high temperature and long processing time in the available methods, such as laser ablation, inverse micelle, solid state techniques, etc. Further, the properties of DMS depend on preparative methods, which are not universal for all DMS. Because of these issues, contradicting results on room temperature ferromagnetism of these materials have been reported in the literature. In addition, the available methods allow synthesis of only simple compound with two elements, such as oxides doped with transition metal ions. However, it is advantages to also produced mixed oxides (three or more elements) as hosts for transition metal and rare-earth ions. Finally, it is more difficult to synthesize DMS quantum dots as compared to bulk DMS. Thus, it is important to develop a novel synthetic method for uniform doping of transition metal and/or rare-earth ions in semiconductors, which allows production of DMS quantum dots based on both simple and mixed oxides.

The aqueous combustion method can be explored to synthesize transition metal ions doped ZnO. However, magnetic measurement reveals absence of ferromagnetism even uniform doping of transition metal ions in ZnO. Color of ZnO:Co^{2+} is blue, which exhibits ferromagnetism. But, the aqueous combustion method yields green color of ZnO:Co^{2+} and this compound exhibits absence of ferromagnetism. Therefore, in addition to uniform doping of transition metal ions in semiconductor, there do have other factors that contribute ferromagnetism. Thus, recent reports on ferromagnetism on cobalt and iron doped $La_{0.5}Sr_{0.5}TiO_{3-x}$ perovskite indicate that ferromagnetism is due to intrinsic one.

INORGANIC PHOSPHORS FOR SOLID STATE LIGHTING

Phosphors find various applications ranging from fluorescent lamp to immunoassay. Among them, fluorescent lamp is largely used and hence, lot of efforts is focused on it. Now, there is an immense research activity in academia and industries in inorganic phosphors after the discovery of blue or near UV LED. The combination of blue LED and yellow phosphor leads to replace the Hg discharge fluorescent lamp. Thus, toxic mercury could be successfully replaced now-a-days. Phosphors in the blue LED convert part of blue emission/near UV of LED into green and red regions, and the combination of these colors results in white light. Nakamura patented yttrium aluminum garnet (YAG:Ce^{3+}) doped with Ce^{3+}, which is the best phosphor known today for obtaining white light from blue LED. Here, the part of blue light is transferred to yellow. Thus, the combination of blue and yellow leads to white light.

The coordination number of Ce^{3+} in YAG is eight. Therefore, 8-coordinated ion in host materials is required for the development of non-YAG materials. Therefore, other host for Ce^{3+} can be thought of YVO_4 where coordination of yttrium in YVO_4 is eight. The general formual of pyrochlore structure is $A_2B_2O_7$. In this compound, A ion occupies eight coordination number, B ion occupies at octahedral coordination (Fig.63) and O ion occupies at tetrahedral coordination.

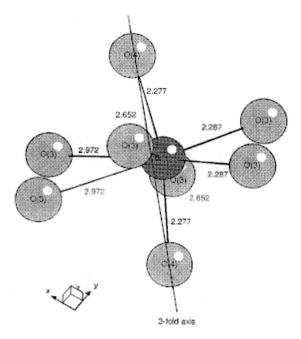

Fig.63: Coordination number view for A^{3+} ion in $A_2B_2O_7$

Thus, this compound is suitable for substituting A ion by Ce^{3+}. Typical examples for pyrochlore structure are $Ln_2Ti_2O_7$, $Ln_2Zr_2O_7$, $Ln_2Sn_2O_7$, $Ca_2Sb_2O_7$ etc. These compounds are potential candidates for Ce^{3+} doping and exhibiting yellow emission under blue LED.

There are two known activators, Ce^{3+} and Eu^{2+} for the LED solid state lighting. These activators in a particular host exhibit required emissions. It is usually observed that Ce^{3+} doped phosphor shows emission about 100-150 nm lower than that of Eu^{2+} emission in the same host. Therefore, it is a thumb rule in the solid state lighting to predict the emission wavelengths for Ce^{3+} and Eu^{2+} activators.

PHOSPHORS FOR PLASMA DISPLAY PANELS

When large screen with flat TV is attracted much interest today market in display companies, it is important to understand the function of it and the phosphors used in it.

There are four types of flat panel displays for a large screen display, and these are display with cathode ray tubes (CRT), liquid crystal display (LCD), electroluminescence display (ELD) and plasma display panel (PDP).

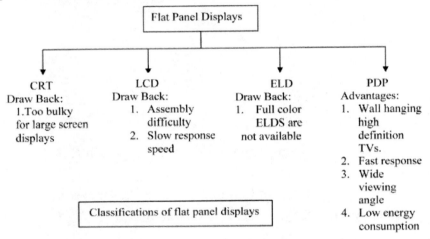

Classifications of flat panel displays

From the flow chart above it is very clear that PDP has several advantages over other displays for large screen display. Therefore, there is a large number of research activities on PDP phosphors to improve the existing phosphors or to discover potential phosphors.

Basic principle of plasma display panel:

This is made up of a glass substrate with two electrodes. Red, green and blue phosphors are coated inside of the glass substrate. Xe or Xe-He gas is filled with glass substrate. During discharge, the plasma produces shorter or higher energy wavelengths such as 130, 147 (the maximum intensity) and 172 nm. Phosphors coated inside the glass substrate convert high energy wavelengths into visible region.

PDP phosphors:

Currently used phosphors in PDP are Eu^{3+} activated Y_2O_3 and $(Y,Gd)BO_3$ red, Mn^{2+} activated Zn_2SiO_4 and $BaAl_{12}O_{19}$ green and Eu^{2+} activated $BaMgAl_{14}O_{23}$ and $BaMgAl_{10}O_{17}$ blue phosphors. Thus, PDP phosphors resemble fluorescent lamp phosphors.

Characteristics of PDP phosphors:

1. Hosts and activators should be stable during fabrication and harsh operations of PDP.
2. Hosts and/or activators should absorb radiations between 130-180 nm and absorbed energy should be transferred into activators to show visible emissions.

Commonly used hosts for doping rare earth or transition metal ions are aluminates, borates, phosphates, silicates and some fluorides. Therefore, it is obvious that main group elements are only explored as hosts to get PDP phosphors. Even though, Mn^{2+} is well known green activator there is not much research on transition elements except Zn element. However, it may be possible to explore VO_4^{3-} group as host for PDP phosphors.

Rare earth ions such Eu^{3+}, Eu^{2+}, Tb^{3+} are explored as dopants in oxide hosts to get PDP phosphors. Table 5 summarizes decay times of commonly used PDP phosphors.

Table 5: Decay time of PDP phosphors

Phosphors	Decay Times, ms
Red: Y_2O_3:Eu	1.3
$(Y,Gd)BO_3$:Eu	4.3
Green: Zn_2SiO_4:Mn	11.9
$BaAl_{12}O_{19}$:Mn	7.1
Blue: $BaMgAl_{10}O_{17}$:Eu	<1
$BaMgAl_{14}O_{23}$:Eu	<1

It is very clear that when the decay times are almost same for blue and red phosphors it is very large for Mn^{2+} activated green phosphors.

Mechanism of PDP phosphors:

It is quite common that the higher energy of VUV radiation is absorbed by host lattice and it is transferred to activators to show visible emissions. In the borates, BO_3^{3-} group absorbs the 150 nm wavelength whereas 150-160 nm absorption in phosphates is due to PO_4^{3-} group. In the case of aluminates, broad excitation band near 175 nm is observed. Some fluorides such as $LiYF_4$, LaF_3 and YF_3 are having 120 nm excitation band.

Even though PDP phosphors resemble fluorescent phosphors, rare-earth doped $NaLaP_2O_7$ and $NaGdP_2O_7$ found useful in PDP can not be used in fluorescent lamps because of Hg vapor reaction with Na ion present in the phosphors.

Research needed:

So far aluminates, silicates, borates, phosphates and fluorides are explored as hosts for VUV absorption. Therefore, it is highly desirable to direct the attention towards combinations of these hosts, namely, borophosphates, borosilicates, aluminoborates, aluminosilicates as hosts for rare-earth ions to convert VUV into visible radiation. Thus, new chemistry such as synthesis, structure, characterization and property can be understood.

PHOSPHOR MATERIALS IN FLUORESCENT LIGHTING

Phosphors are mostly defined as inorganic crystalline materials either doped with activator alone or simultaneous doping of activator and sensitizer. There is also exception in this definition of phosphors. Literature survey reveals single component material with 100% host as a good luminescence. Phosphors are mainly emitting in the visible region. But, excitation varies from gamma-rays to infrared to mechanical or chemical energy. In the case of activator alone doping, luminescence process is mostly involved directly with activator alone. i.e. absorption, excitation and emission all in phosphors take place with dopant, activator. Typical example for this case is Eu^{3+} ion doped Y_2O_3 in low pressure Hg lamp application. In this system, Eu^{3+} absorbs 254 nm wavelength arises from low pressure Hg fluorescent lamp by a charge transfer from Eu^{3+} to oxide ion. Then, it shows intense emission in the visible region. The luminescence process involving activator ion alone is represented below in Fig.64.

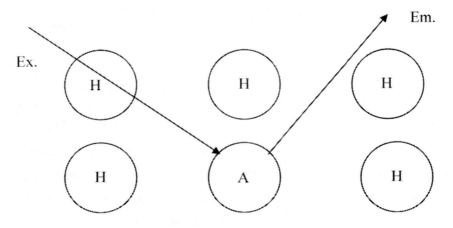

Where H = Host and A = Activator
Ex.=Excitation & Em.=Emission

Fig. 64: Direct involvement of activator ion in the Excitation and Emission Processes

Therefore, the main goal of host in this category is to keep activator in its environment and to allow the activator to involve in the luminescence processes.

In the second case, sensitizer is used in addition to activator ion doping since the doped activator ion or host does not absorb a particular excitation wavelength. Therefore, sensitizer absorbs excitation wavelength and emits in the visible or ultraviolet region. Part of emission from sensitizer is transferred to activator and the activator shows intense visible emission in addition to sensitizer emission. Fig. 65 shows the use of sensitizer in phosphors.

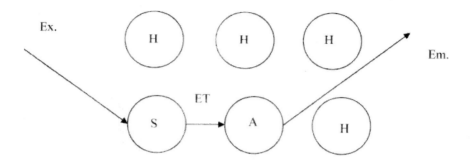

Where S = Sensitizer, A = Activator, H = Host, Ex. = Excitation
Em. = Emission and ET = Energy transfer

Fig.65: Role of sensitizer to activate activator ion in phosphor

It is also important to mention about energy migration through host from sensitizer to activator in phosphors. This process often leads to lower the concentration of sensitizer and/or activator. One important example for this case is Gd-based phosphors. The use of sensitizer is understood from a classical phosphor, $Ca_5(PO_4)_3(F,Cl):Sb^{3+},Mn^{2+}$. In this system, Mn^{2+} ion is activator which does not absorb 254 nm in low pressure Hg fluorescent lamp. Therefore, antimony ion is also doped in calcium halophosphate along with Mn^{2+} ion. Sb^{3+} ion absorbs 254 nm. Here, antimony ion transfers its excited electron into Mn^{2+} ion excited state and the excited electrons in Mn^{2+} decays to ground state by emission in yellow region. The combination of blue emission from Sb^{3+} ion sensitizer and yellow emission from Mn^{2+} ion activator leads to white light in low pressure mercury lamp.

Configurational Coordinate Model:

Configurational coordinate model is a simple model to explain excitation and emission processes. Thus, it considers "U" shape for electronic energy levels, where as, vibrational energy levels are expressed by horizontal lines. Fig. 66 shows excited and ground states for electronic and vibrational energy levels.

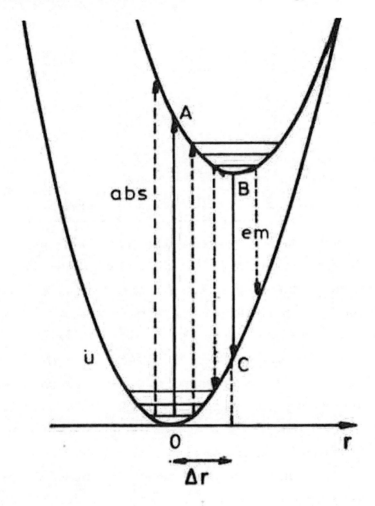

Fig.66: Configurational coordinate model to explain phosphor properties

During excitation process from electronic ground state to electronic excited state there is no change in vibrational modes. Thus, dopant in the vibrational ground state from ground state electronic level is excited to higher vibrational levels in first excited electronic energy level. Then, dopant relaxes in the electronic excited state from higher vibrational levels to ground state vibrational level. During this process, dopant dissipates the energy difference as heat. Now, it is observed visible luminescence of dopant when it returns to ground state electronic level, but, at higher vibrational levels. Then, vibration relaxation at ground state electronic level of dopant returns to ground state vibrational level of ground state electronic level. Because of vibrational relaxation of dopants, emission wavelength is always greater than that of excitation wavelength.

Term Symbols:

Term symbols are used to represent energy levels of dopants in phosphors. They are obtained by spin-orbit coupling employing Hund's rule and Pauli Exclusion Principle. A term symbol is represented as

$$^{2S+1}L_J$$

Where $2S+1$ = multiplicity

L = Total orbital angular momentum

J = Total angular momentum.

For different "L" values, the term symbol is different. Thus,

For $l = 0$, the term symbol is S

For $l = 1$, the term symbol is P

For $l = 2$, the term symbol is D

For $l = 3$, the term symbol is F and so on.

Term symbol obtained for $4f^1$ electronic configuration is explained below now.

↑						

For 'f' orbital, there are seven orbitals with seven values of 'l' from -3 to +3. For one electron the values for 'l' and 's' are 3 and ½ respectively. J values are obtained as follows.

l+s, l+s-1, l+s-2, --- l-s

For $4f^1$, the J values are

3+1/2, 3+1/2-1, 3+1/2-2, 3+1/2-3

Therefore,

       7/2, 5/2, 3/2, ½

s = ½ therefore, 2S+1 = 2

l = 3, therefore, L = F

Therefore, there are four energy levels possible for l = 3 and s = ½ with four different term symbols.

Order of arrangement of energy levels with term symbols:
1. The largest multiplicity (2S+1) has ther lowest energy level.
2. For the same multiplicity (2S+1), the largest "L" has the lowest energy level.
3. For the same multiplicity and L, the lowest J value has lowest energy level.

By employing above three rules, the energy levels for l = 3 and s = ½ are arranged in increasing order as follows.

$^2F_{1/2} < {}^2F_{3/2} < {}^2F_{5/2} < {}^2F_{7/2}$.

Chemistry of doping:

      There are two aspects considered in the chemistry of doping to get the maximum luminescence in the visible regions. One is charge effect and another is size effect and these two effects are described below one by one.

Charge effect:

      It is important to avoid aliovalent doping in phosphor hosts. i.e. charge of dopants should be same as that of charge of host element at which doping takes place. This is essential to avoid nonstoichiometry formation in the doping process. Nonstiochiometry, some time, may lead to formation of centers for nonradiative decay. This is understood by considering doping of Eu^{3+} ion in ZrO_2 host. Here, the charge of Eu^{3+} ion is different from the charge of Zr^{4+} ion and hence, ZrO_2 is not an ideal host for Eu^{3+} doping to get red luminescence.

Size effect:

Another interesting effect observed in the chemistry of doping is size effect. i.e. size of dopants is almost similar to that of dopant sites. Otherwise, it is not possible to dope up to the level of required concentration of dopants. This is very easily understood by considering Eu^{3+} doping in gamma alumina to get red luminescence. Because of dissimilar in ionic sizes of Eu^{3+} and Al^{3+}, very little quantity of Eu^{3+} ion can be doped at Al^{3+} site in gamma alumina. In order to enhance emission intensity, it is not possible to increase the concentration of Eu^{3+} doping in gamma alumina. Sometime, instead of Eu^{3+} doping at higher concentration, $EuAlO_3$ perovskite phase starts to forming while increase the concentration of Eu^{3+} in the doping process.

Low pressure Hg based fluorescent lamp:

This type of fluorescent lamps is widely used now-a-days. This is made up of glass tube sealed two ends as shown in Fig.67..

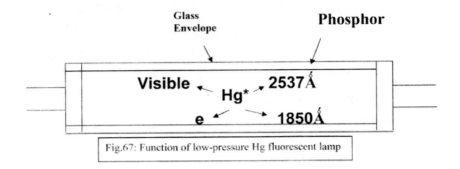

Fig.67: Function of low-pressure Hg fluorescent lamp

Inside the glass tube, noble gas and mercury are present at 400 and 0.8 Pa respectively. During the electric discharge, mercury atoms present in the glass tube as a vapor get excited to higher electronic energy levels. Because of low pressure of mercury vapor, the emissions from excited states show atomic line emissions. Thus, they mainly emit 85% of 254 nm, 12% of 185 nm and 3% of them are in the visible regions. Phosphor materials are coated inside wall of the glass tube. The thickness of the phosphor materials vary from 20 to 40 microns. The phosphor materials essentially absorb 254 and 185 nm wavelengths and they emit in the visible regions to get white light. The phosphor materials are very limited since they have to convert UV light of 254 and 185 nm wavelengths into visible regions. There are also phosphor materials further limited by reaction of potential phosphors with mercury. Therefore, stable and refractory or inert phosphor materials are explored as luminescent phosphors in Hg based fluorescent lamps.

Luminescence property of phosphors:

Rare-earth activators:

Phosphors activated by rare-earth ions exhibit some peculiarities. In the energy level diagram of rare-earth ions, luminescence processes often correspond to electronic transitions within the incompletely filled 4f shell, which is extensively shielded by the crystal field-splitting that results from the incorporation of rare-earth ions into a host lattice. Consequently, these phosphors have narrow band spectra which are to a great extent independent of the nature of the host lattice. Because of the low interaction with the crystal lattice, the luminescence quantum yield of phosphors activated with rare earth ions is often higher compared to other phosphors. Quenching occurs only at higher temperatures or higher activator concentration. The energy level diagrams of Ce^{3+}, Eu^{3+}, Eu^{2+} and Tb^{3+} are given in Fig. 68.

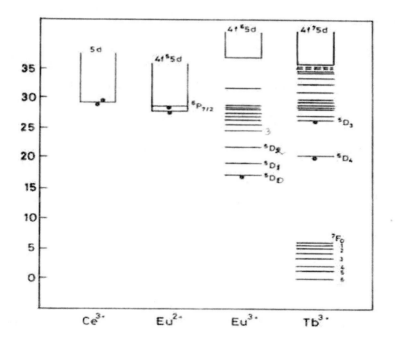

Fig.68: Energy level diagrams of the most commonly used rare-earth activators

Energy in the y-axis is given as 10^3 cm^{-1}

CARBONITRIDE WITH SP³ CARBON

Theory predicted that sigma bonded carbonitride should be superharder than diamond and the nominal composition of carbonitride is C_3N_4. This carbonitride resembles with Si_3N_4 but there are several phases have been observed in carbon-nitrogen system. These phases are alpha, beta, cubic and pseudocubic C_3N_4. But, there is no report so far to get cubic C_3N_4 with sigma bonded carbon in bulk synthesis to prove the theoretical predictions. However, there are a few articles including US patent (6,428,762) on amorphous C_3N_4 synthesis with various morphologies. The typical synthesis involves reaction between cyanuric chloride ($C_3N_3Cl_3$) and Li_3N as a nitridation in a closed container (free from air) at 380°C for about 4 hours. The formation of C_3N_4 is given below in an equation.

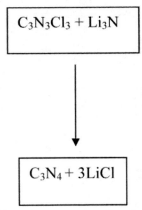

The driving force for this reaction is due to formation of thermodynamically stable LiCl like observed in solid state metathesis reaction.

Research needed:

So far starting materials used to synthesize C_3N_4 contain SP² carbon. Therefore, carbon has both sigma and pi bonds in the final product, C_3N_4. In order to get SP³ carbon in the final product, C_3N_4, SP³ carbon in the starting material is required. To achieve this, hydrazine based compounds can be considered. Thus, triformyl trisazine (TFTA, $C_4N_6H_{12}$) can be a suitable precursor to get covalently bonded C_3N_4 since TFTA has SP^3 carbon.

The carbon to nitrogen ratio in TFTA is 4.5, which is higher than that of C_3N_4 compound. Once the precursor is chosen, the next step is the experimental design to get sigma bonded C_3N_4 followed by various characterizations.

There are two ways by which TFTA precursor can be used to get covalently bonded crystalline C_3N_4 and these ways are described below.

(i) Thermochemical decomposition of TFTA:

TFTA contains both carbon and nitrogen with covalent bonds. Therefore, it is a good precursor to synthesize crystalline and covalent bonded C_3N_4. In principle, on heating in air, TFTA completely decomposes into gaseous products. In order to get C_3N_4 solid as a product, TFTA should be heated in the flow of ammonia and/or nitrogen gas (without oxygen). Thus, it is possible to get crystalline C_3N_4 with SP^3 carbon.

(ii) Solvothermal synthesis of C_3N_4

In this method, TFTA is subjected to decompose in benzene solvent under autogeneous pressure in the temperature range 150-250°C. In order to accelerate this reaction, zinc metal powder might be included in the solvothermal reaction. Formation of C_3N_4 from TFTA under this condition is represented below.

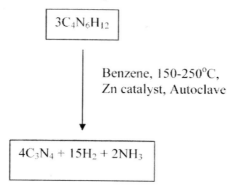

Structure of covalently bonded C_3N_4:

Solid compounds from the first row lighter elements with covalent bonding are considered as harder than diamond. This prediction was published in 80's. Therefore, this structure of this compound is assumed to very similar to Si_3N_4. Four covalently bonded Si with nitrogen is replaced by carbon and the structure of C_3N_4 can be shown below (Fig. 69).

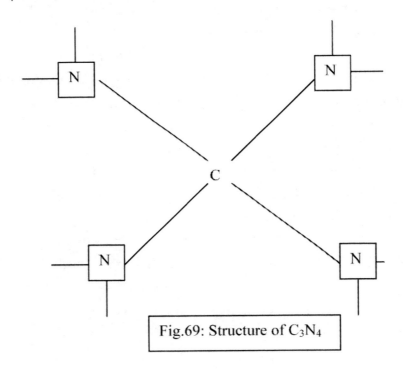

Fig.69: Structure of C_3N_4

OPEN FRAMEWORK MATERIALS

There has now been a tremendous research work in the area of organically templated open framework materials. The open framework materials differ from thermodynamically stable condensed materials, and ring structures that usually observed in open framework materials are absent in the condensed materials. These open framework materials resemble well-developed zeoletic materials. The zeoletic materials are usually aluminosilicates. The differences between zeolites and organically templated open framework materials are summarized in the following table 7.

Table 7: comparison of properties of zeolites with open framework materials

Zeolite	Open Framework Material
Alumino Silicates	Transition metals also present
Acid Catalysts	Could be redox catalysts
Tetrahedral frameworks	Can be 5- or 6- coordination in addition to tetrahedral
Thermally stables	Not with all the compounds
Diamagnetic materials	Can be magnetic materials
Non-luminescent	Can be luminescent materials
Three-dimensional pores	1- or 2- or 3-dimensional pores

The stability of organically templated open framework materials is questionable. Templated organic molecules form extensive hydrogen bonding with framework oxide or hydroxyl group. Therefore, calcination destroys the hydrogen bonding, and open framework materials lead to the formation of stable condensed materials. For example, except a few compounds in organically templated metal phosphates, all the compounds loose the thermal stability.

Combination of open framework materials with semiconducting property may lead to removal of organic molecules occluded in the pores of open framework by photochemically without destroying the structures. Thus, protonated form of the compound is obtained and this type of conversion is represented below (Fig. 70).

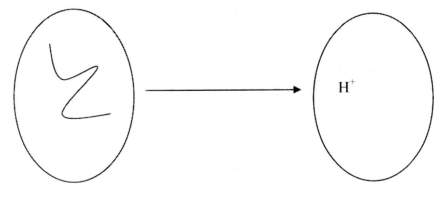

Fig.70: Formation of protonated open framework material

The protonated form could be
        1. better photocatalysts for photosplittings of water
        2. proton conductors.

THREE WAY CATALYSTS

In a chemical reaction, conversion of reactants into products takes place through a energy barrier (transition state) and the height of the barrier is represented by $e^{-Ea/RT}$ where Ea is called activation energy, T is the temperature and R is the gas constant. It is represented by the following diagram (Fig.71).

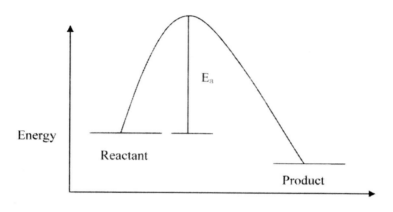

Fig.71: Representation of a reaction using activation energy.

Reaction can be accelerated by two ways. One way is by increasing temperature. Another way is by using catalyst, which actually decreases Ea by lowering the transition state. This is explained in the diagram (Fig.72). Ea' is lower than that of Ea since catalyst reduces activation energy.

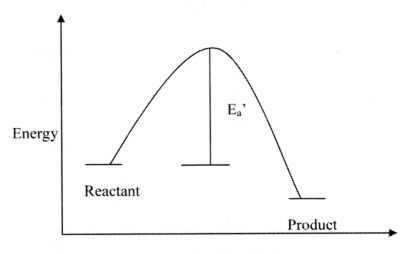

Fig. 72: Effect of catalyst on a reaction

In three way catalysis, the following reactions take place in an automobile exhaust.

Oxidation:
$$2CO + O_2 \rightarrow CO_2$$
$$HC + O_2 \rightarrow CO_2 + H_2O$$

Reduction:
$$2CO + 2NO \rightarrow 2CO_2 + N_2$$
$$HC + NO \rightarrow CO_2 + H_2O + N_2$$

where HC=unburned hydrocarbon

In order to achieve both the oxidation and reduction reactions, catalyst should be bi-functional, i.e. catalyst oxidizes CO and unburned hydrocarbons into CO_2 and H_2O and reduces NO into N_2 and H_2O. Conventional catalyst is Pd dispersed Al_2O_3 or CeO_2-ZrO_2 composite. The main drawback of the conventional catalyst is that particle growth occurs at such a high temperature in the automobile exhaust. Therefore, activity is decreased due to lowering in surface area. Of late, Pd-perovskite was found to be self-regeneration for automotive emissions control. The intelligent catalyst is $LaFe_{0.57}Co_{0.38}Pd_{0.05}O_3$. The mechanism of the intelligent catalyst in automotive emissions control is as follows. During oxidation and reduction atmospheres structural changes takes place in the intelligent catalyst. Thus, during reduction, Pd is getting released from the intelligent catalyst and Pd gets into the perovskite catalyst in oxidation atmosphere. At high temperature in automobile exhaust, perovskite catalyst undergoes such changes and thus, particle size is not affected and maintains catalytic activity unlike observed particle growth in the conventional catalyst.

Research Needed:

The above presentation clearly demonstrates that catalyst should accommodate Pd during oxidation process to avoid Pd catalyst growth. Or in other words, during oxidation process, the high temperature should be used for Pd incorporation in its oxide catalyst rather than particle growth. Therefore, the research needed in this direction is to select catalyst that accommodates Pd in its oxide form as well as the selected material should be catalyst for these types of reactions.

FUEL CELL MATERIALS

Fuel cells operating at a range of temperatures will play an important role in the next generation of electricity production with zero pollution. Among the various fuel cells, solid oxide fuel cell (SOFC) and polymer electrolyte membrane (PEM) fuel cell are important to mention now. First, a fuel cell function is outlined and then, the main differences of the two fuel cells are highlighted.

Function of a fuel cell:

A fuel cell consists of anode, cathode and electrolyte. At the anode, fuel such as hydrogen gas is oxidized into proton and electron whereas at the cathode, oxygen is reduced into O^{2-}, which then combines with proton to yield water as gas. The electrons released at the anode passes through external wire to cathode and thus, electricity is produced. Function of a fuel cell is schematically described in Fig. 73.

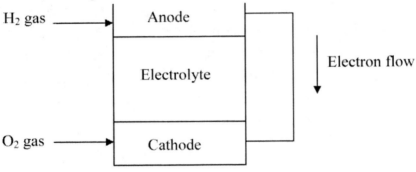

Fig. 73: Schematic diagram of a fuel cell

Brief description of PEM fuel cell:

Anode and cathode are platinum metals. Electrolyte is a polymer membrane. PEM fuel cell operates at 70°C and the main function of polymer electrolyte in the PEM fuel cell is to carry proton from anode to cathode, wherein it combines with oxide ion to give water. Polymer membrane used currently is not stable beyond 100°C and hence, it is not possible to accelerate proton transportation from anode to cathode. Therefore, it is important to consider inorganic materials, according to me, containing proton as electrolyte. These inorganic materials might be stable enough to accelerate proton transportation from anode to cathode. Such examples are $H_4Nb_6O_{17}$, $H_2Ti_4O_9$, $HCa_2Nb_3O_{11}$ and $HNbWO_6$. Electrolyte should not allow the gases H_2 and O_2 to pass through it and there is another branch of materials science arises here to study the sintering property of protonated inorganic materials. Sintering is the process by which polycrystalline materials packed very closely such that polycrystalline materials achieve near 100% density of single crystal density. This is usually done below melting or decomposition point of the material. A schematic diagram showing function of PEM fuel cell is found in Fig. 74.

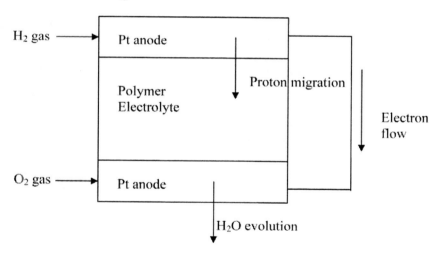

Fig. 74: Function of a PEM fuel cell

Brief description of SOFC:

In this fuel cell, anode is cermet and cathode is oxide. Electrolyte employed is oxides. The operating temperature is greater than 700°C. At this temperature oxide electrolyte transports oxide ion formed at cathode to anode where in oxide ion combines with proton to yield water. The required research in this topic is to reduce the operating temperature of SOFC. This is possible by investigating oxide ion transportation using various oxides. Fig. 75 shows a function of SOFC.

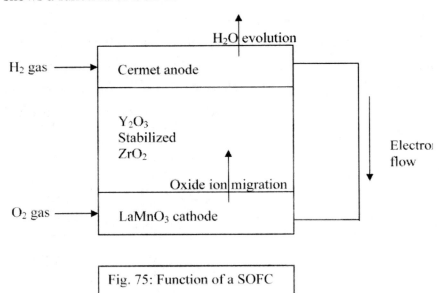

Fig. 75: Function of a SOFC

Differences between PEM fuel cell and SOFC:

Table 8 compares the main differences between PEM fuel cell and SOFC

Table 8: Comparison of PEM fuel cell and SOFC

PEM Fuel Cell	SOFC
Operating Temperature is ~80°C	>700°C
Polymer electrolyte	Oxide ion conductor as an electrolyte
Electrolyte carries proton from anode to cathode	Electrolyte carries oxide ion from cathode to anode
H_2O is formed at the cathode	H_2O and CO_2 are produced at anode
Electrodes are Pt	Anode is cermet and cathode is oxides

THERMOELECTRIC MATERIALS

The title itself implies that this material deals with thermal and electrical energies. Except material synthesis, all the properties are physics oriented. As a chemist I try to explain it very briefly in a simplified manner. Thermoelectric material is a semiconductor that finds applications in thermoelectric devices. These devices, making use of property of semiconductor, is simple, free from noise pollution and no toxic or no green house gases evolve. Thermoelectric device converts thermal energy into electrical energy and as well as it cools one end when it is connected the other end to external input (Refrigeration).

Construction of Thermoelectric Device:

This device makes use of two different semiconductors, one is n-type and another is p-type semiconductors, connected in series. N-type semiconductor carries current by electron whereas p-type semiconductor carries current by hole. There are two applications found using these properties of semiconductors.

In a refrigeration, one end of two types of semiconductor is connected to external power. Charge carriers carry current as well as heat. Thus, external power attracts the charge carriers towards it. Because of movement of charge carriers towards external power, action of cooling takes place at the other end. This is due to carrying of heat by charge carriers as shown in the following figure (Fig.76).

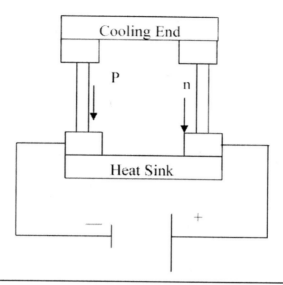

Fig.76: Action of refrigeration using thermoelectric materials

When one end of two semiconductors is heated, voltage develops across the two semiconductors. Thus, heat is converted into electricity. When one end is heated, charge carriers move towards heating end. The following figure represents the function of power generation using thermoelectric materials (Fig.77).

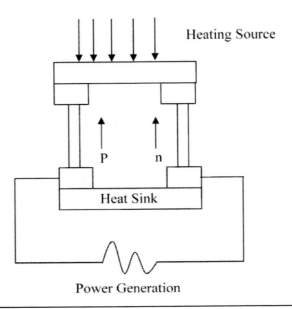

Fig.77: Function of power generation using thermoelectric materials

The efficiency of thermoelectric materials is determined by ZT where T= temperature and Z is a figure of merit. Z depends upon the electrical and thermal properties of semiconductors. To be a better semiconductor, they should show the lowest thermal conductivity and the highest electrical conductivity. Semiconductors having heavier elements are used in these applications since the thermal conductivity is lower for them. Recently, oxide materials are also found to show better property in these applications. A thermoelectric refrigerator requires ZT = 3 at room temperature.

BIOCERAMICS

Bioceramics are mainly inorganic oxide materials and these find applications in replacing damaged and diseased part of the body. These are classified into two broad areas. One of them is bioinnert materials that do not have any interactions with human body. Examples for this type of materials are α-alumina and zirconia. The another type of them is bioactive materials and they do have interactions with tissues in the body. Typical examples are calcium phosphate based oxides.

For the actual clinic applications, these bioceramics with good mechanical properties and chemical stability are required. For obtaining better mechanical property of these material, fine grain sintered body is prerequisite. Therefore, nanomaterials are required in this topic to get sinter-active materials.

It is also important to mention here that porous bioceramics are good for bone in-growth. However, macroporous materials as bioceramics are required but good mechanical property can not be obtained with such a large porous. Therefore, in order to improve mechanical property, a seconday phase such as Al_2O_3, TiO_2, SiO_2, ZrO_2 is required.

It is very clear from the literature survey on bioceramics that there are two directions research activity in bioceramics field being carried out. One direction is the fabrication of fine grained sintered materials that find application in load bearing body parts. Another direction is the fabrication of porous materials that find application in tissue growth in bones.

Research needed:

To get sinter-active bioinnert materials, nanomaterials are required. Thus, it is possible to get fine grain sinter-active materials such as Al_2O_3, ZrO_2 by hydrothermal synthesis. To synthesize Al_2O_3

nanomaterial, for example, aluminum powder should be treated with oxidizing agent in the autogeneous condition. Under this condition formation of nanomaterial is possible and hence, it leads to fine grain sinter body.

Hydrothermal synthesis could be extended to synthesize porous hydroxyapatite using organic templates.

THE HYDROGEN ECONOMY: THE FOREVER FUEL, H_2

Hydrogen gas is considered as a forever fuel because of availability of hydrogen in different forms. Therefore, research in America and rest of the world is directed to generate, store and applications of hydrogen gas. To explore hydrogen gas as a fuel, either internal combustion engine or fuel cell is required to get energy from a straight forward reaction between hydrogen and oxygen gases. This reaction yields only water vapor. The ultimate goal in hydrogen economy is to replace currently available dependent and carbon-based fuel by independent hydrogen gas fuel. This replacement is also proved not to produce green house gas such as CO_2.

By 2020, the world will face an oil crisis. The majority of American automobile owners will suffer greatly from this crisis if new energy sources are not available. The first oil shortage in the 1970s and 1980s was economically and politically induced. This time, however, the crisis will be based on a real shortage of oil for fuel. Although optimist argue that new oil fields like exist 3280 feet or more below the surface of the oceans, the process of finding and obtaining it is very expensive and the technologies to do so are not fully developed. Therefore, it is highly desirable to avoid dependence of fossil fuel. Also, carbon-based oil produces carbon dioxide (CO_2) as a byproduct that will likely increase global temperature by 2.52 to 10.44°F over the next one hundred years. These temperature increased will cause the polar ice caps to melt. Therefore, alternative to carbon based oil fuels must be researched. This could be achieved by using Hydrogen (H_2) as a fuel that produces only healthy water vapor when it burns, and does not heat up the air.

Chemical reactions could be explored to produce hydrogen gas. One such example is the violent reaction of reactive metal with acid. Thus, Zn reaction with HCl yields $ZnCl_2$ and H_2 gas. Similarly, Fe metal reacts with sulfuric acid leads to production of H_2 gas and $FeSO_4$. These reactions are represented as follows.

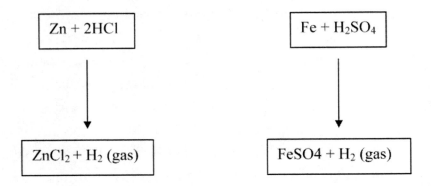

Another chemical reaction is that metals such as K, Na and Ca react vigorously with water producing enough heat to ignite the hydrogen. Example of Na reaction with water is represented by the following reaction.

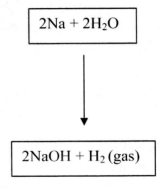

The formed NaOH in the above mentioned reaction further can be treated with Al or Si in presence of water to produce H_2 gas.

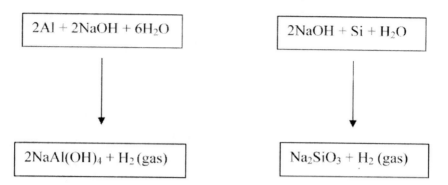

Hydrides could be explored as a source for hydrogen. Calcium hydride or sodium borohydride is treated with water to produce hydrogen gas and the reactions involved in these processes are represented below.

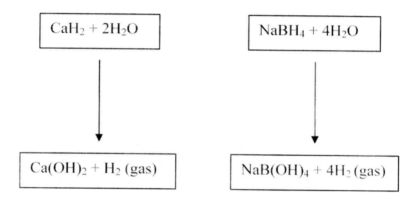

RESEARCH NEEDED IN PHOTOSPLITTING OF WATER

IGNEOUS COMBUSTION SYNTHESIS OF OXIDE SEMICONDUCTORS AND THEIR USE FOR THE PHOTOSPLITTING OF WATER AT ROOM TEMPERATURE

Background

Solar-splitting of water using semiconductors plays a vital role in photo catalysis field. Using photon to split water into H_2 and O_2 is aimed at harness the solar energy because solar radiation reaching at the earth surface for 1 h is equivalent to fossil energy consumed by the world for 1 year. Therefore, research on photo catalytic breakdown of water is of world wide interest. The reduction of water into H_2 and oxidation of water into O_2 using semiconductors are represented below.

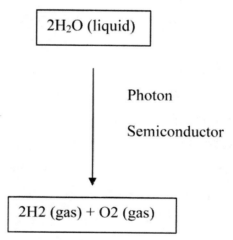

Fig. 78 represents energy profile diagram for the photosplitting of water. It is very clear that potential of product is greater than that of reactant and hence, it is a reversible reaction. In order to forward the reaction, products should be removed from the reactor continuously.

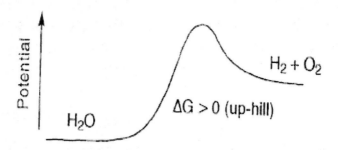

Fig.78: Energy profile for water splitting

The principle of photo splitting of water into hydrogen and oxygen using semiconductors with appropriate band gap and positions of valence and conductance bands is represented in Fig. 53. The energy difference for the redox reaction of H_2O into H_2 and O_2 is 1.23 eV. Therefore, semiconductor with band gap greater than 1.23 eV is required for the solar-splitting of water into H_2 and O_2. Also, this photo evolution of gases from water depends upon the position of conduction and valence bands of semiconductors. Thus, the bottom level of the conduction band has to be more negative than the redox potential of H^+/H_2 (0 eV Vs NHE), while the top level of the valence band be more positive than the redox potential of O_2/H_2O (1.23 eV). It indicates that electron and hole are generated under photons with energy equal or greater than the band gap of semiconductor. Thus, electrons are promoted to the conductance band whereas holes remain in the valence band. These charge carriers are responsible for photo splitting of water. In a real situation, there are two main things occurring in a semiconductor particle and this is represented in the Fig. 79 below.

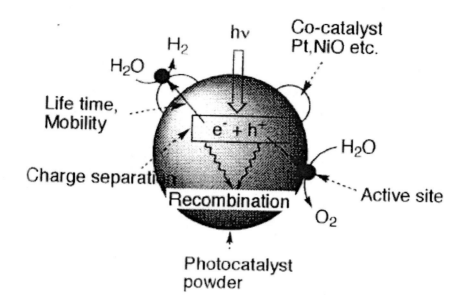

Fig. 79: Photosplitting of water using particle semiconductor

The desire thing is the charge separation of electron and hole and migragtion of them to the surface of the particle. In order to improve the catalytic activity, a co-catalyst such as Pt or NiO is coated at the surface of the particle. An unwanted process is recombination of charge carriers, which leads to a lower yield for the reaction.

Among the semiconductors, TiO_2 is widely studied for photo splitting of water. TiO_2 semiconductor shows photo splitting of water into H_2 and O_2 under UV irradiation. Similarly, another system studied extensively is $K_4Nb_6O_{17}$. The success of $K_4Nb_6O_{17}$ semiconductor in the photo splitting of water is due to the fact that hydrogen evolution and oxygen evolution take place in different interlayer space and hence, recombination of the gases is avoided. However, none of the layered compounds exhibited photo splitting of water into H_2 and O_2 under visible irradiation. On the other hand, Pt/CdS is one of the photo catalysts that show evolution of H_2 gas using sacrificial reagent under visible irradiation, while WO_3 or $BiVO_4$ shows evolution of O_2 gas under the same conditions using different sacrificial reagent.

Aims

a) The forgotten research in solar-splitting of water is use of bulk/nano/transition metal ions doped semiconductors as photo catalysts even though there was a report on complete splitting of water into O_2 and H_2 under visible irradiation using nickel doped $InTaO_3$ semiconducting oxide. Therefore, now is the time to devote research activities toward bulk/nano semiconductors with or without doping transition metal ions.

a) The other direction forgotten in this field is use of mixed oxide semiconductors. Of interest is $BiVO_4$ (mixed oxide semiconductor) which shows excellent and clean O_2 production in presence of sacrificial reagent. It is also interesting to notice that Bi_2O_3 is a semiconductor and V_2O_5 is another semiconductor. Combination of these two individual semiconductor leads to mixed oxide semiconductor. Therefore, I assume the following thumb rule.

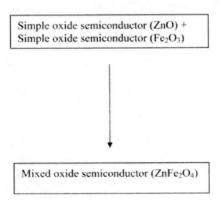

Thus, the combination of two different semiconductors such as ZnO and Fe_2O_3, leading to $ZnFe_2O_4$ might be suitable for photo splitting of water.

Similarly, semiconductor rich mixed oxides could be potential semiconductors for photo splitting of water into H_2 and O_2.

c) Among the known various semiconductors, SnO_2 and WO_3 (Fig.80) have their lowest valence bands the most positive than that of redox potential of H_2O into O_2.

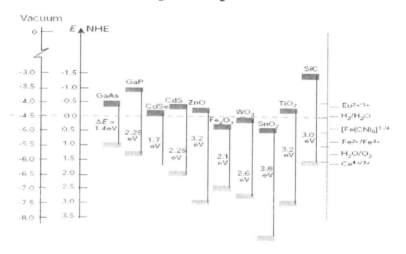

Fig.80: Energy levels of semiconductors

Therefore, SnO_2 and WO_3 could be potential candidates for photo splitting of water into O_2 in the presence of sacrificial reagent. It is noted that these semiconductors can not produce H_2 from water under photon since their lower level of conduction band is less negative than that of redox potential of H_2O into H_2. Therefore, I would like to couple two complementary semiconductors as nano composites to photo-splitting of water into H_2 and O_2 without any sacrificial reagent. Thus, WO_3 or SnO_2 oxide will be incorporated into the layer of $H_4Nb_6O_{17}$ semiconductor, which is known to produce H_2 from water and the nanocomposites of $H_4Nb_6O_{17}/WO_3$ and $H_4Nb_6O_{17}/SnO2$ will be efficient for complete splitting of water into H2 and O2 under photon.

Approach

Now, I propose the following semiconductors (tables 9 & 10) for the solar-splitting of H_2O into H_2 and/or O_2 under visible irradiation.

Table 9: Possible semiconductors with band gap in the visible range

Semiconductor	Band Gap, Ev
CdO	2.00 – 2.20
In_2O_3	2.6 – 2.70
Cd_2SnO_4	2.0 – 2.20
$CdIn_2O_4$	2.7 – 3.00

Table 10 is generated based on thumb rule as stated in the aim section.

Simple Semiconductor	Simple Semiconductor	Mixed Semiconductor or semiconductor rich oxide
ZnO	Fe_2O_3	$ZnFe_2O_4$
ZnO	SnO_2	Zn_2SnO_4
ZnO	In_2O_3	$Zn_2In_2O_5$
BaO (insulator)	TiO_2	$Ba_2Ti_9O_{20}$
BaO (insulator)	Nb_2O_5	$BaNb_2O_6$

In this section, already known semicondutors with band gap in the visible region (Table 9) and mixed oxide semiconductors

(Table 10) are considered to explore as photocatalysts for water splitting.

Synthesis

Igenous combustion method could be employed to synthesize the above mentioned compounds. This synthesis method is a wet-chemical method that occurs with evolution of heat and light. Thus, when rapidly heating aqueous solution of corresponding metal nitrates and fuel (glycine or urea or hydrazine based compounds) in the temperature range 350-500°C the solution boils, froths and catches fire. The in-situ temperature increases to about 1000°C. Such a high temperature is sufficient enough for the formation and crystallization of oxides.

Advantages of Aqueous Combustion Method

(i) Molecular level homogeneity of metal ions is achieved.

(i) Very low ignition temperature (350°C-500°C) is sufficient for getting high "in situ" temperature (>1000°C).

(i) Exothermicity is used for formation and crystallization of oxides

(ii) The entire combustion reaction lasts for less than 5 min.

(i) Combustion method is simple, fast and efficient.

Other synthetic methods employed to synthesize semiconductors include ion-exchange, intercalation and hydrothermal. Powder XRD, DRS, TGA-DTA, chemical analysis will be used to characterize semiconductors. Photo-fission of water into H_2 and O_2 will be evaluated by specially designed experimental set-up.

PHOTOSPLITTING OF WATER USING NANOCOMPOSITE SEMICONDUCTORS WITH OPEN FRAMEWORK STRUCTURE

Background:

The search for new compounds in open framework materials (also called organically templated inorganic materials) stems from the fact that they exhibit wide variety of solid state structures and also equally potential for industrial applications such as ion-exchange, molecular sieve and catalysis. After the discovery of aluminium phosphate there has been extensive work to synthesize phosphates containing main group and transition metal ions. In addition to tetrahedral geometry, there are octahedral and five coordination geometries possible in the transition metal phosphates. Therefore, there are lots of combinations of these coordination geometries.

Since boron finds place in the periodic table just above aluminium there has been effort to synthesize open framework materials containing boron. The interest in boron based compounds arises due to excellent catalytic activity of bulk BPO_4. But, it is obvious that tetrahedral aluminium in aluminium silicate/phosphate can not easily be replaced by boron since boron prefers three fold coordination. However, there are minerals like datolite known to have tetrahedral boron in their compounds. Therefore, lot of efforts is needed to find out a suitable reaction conditions for obtaining boron in tetrahedral coordination. These materials are usually made under hydrothermal conditions containing alkali metal sodium silicate/phosphoric acid and sodium aluminate under basic conditions. Later the alkai metal ions have been replaced by organic amines. The organic amines act as templates or structure directing agents for the formation of these materials. The organic diamines as cations occupy in channels (1-dimensional or 2-dimensional or 3-dimensional) and exhibit hydrogen bonding with framework oxygen/hydroxyl group.

Aim:

Semiconductors with open framework structures and 1-D or 2-D or 3-D pores are not explored in the field of photo splitting of water into H_2 and O_2. Therefore, research in this new and novel direction is removal of organic molecules from inorganic/organic hybrid materials without affecting open framework structure either by calcinations at low temperature <200°C or by photodecomposition at near temperature to get high surface area porous materials.. It is aimed to study photo splitting of water using these porous materials. Nanocomposite semiconductors also will be designed to improve the rate of photo splitting of water.

Approach:

(i) **Development of inorganic/organic hybrid materials with semiconducting property.**

Removal of organic diamines from hybrid materials at above 500°C by calcination leads to destruction of open framework of these materials. Thus, this calcination process converts open framework materials into dense materials and high surface area can not be retained in the dense materials. Therefore, it is aimed to expel organic molecules without affecting the open framework by photodecomposition. If open framework materials possess semoconducting property, it would be a judicious choice for easy removal of organic molecules by UV or visible irradiation . Once, organic molecules are removed by photolysis the protonated form of open framework compound, having higher specific surface area, would be the final product. The protonated form of semiconductors with open framework is potential for photosplitting of water like $H_4Nb_6O_{17}$ semiconductor.

(ii) Exploration of hydrazine as a template in the synthesis of open framework materials.

So far, except one report, organic diamines are employed as a template in the synthesis of open framework materials. The one report is from my work with Prof. Slavi C Sevov at Notre Dame University and the template molecule employed was hydrazine i.e. carbon-free template molecule for the synthesis of cobalt phosphate. Hydrazine in cobalt phosphate donates one nitrogen for a coordination to cobalt ion. Therefore, removal of hydrazine without affecting the structure has been failed in cobalt phosphate system.

Usually, hydrazine based compounds have lower decomposition temperature compare to that of ammonia based compounds. Therefore, hydrazine based open framework materials may possess lower decomposition temperature and hence, protonated open framework materials could be obtained by calcinations at <200°C.

The other way of removal of organic molecule from inorganic/organic hybrid materials is possibly by starting with mixed valence metal phosphate. Calcination of mixed valence metal phosphate converts lower valence metal ion into higher valence since organic cation is expelled out from the structure. Mixed valence metal phosphate could be synthesized by hydrothermal oxidation of transition metal powder with phosphoric acid in presence of organic diamine molecules.

(iii) Nanocomposite Semiconductors for Photosplitting of water

Charge carriers are formed in semiconductor under UV or Visible irradiation, and these charge carriers are responsible for photosplitting of water. In order to have high rate of photosplitting of water, charge carriers have to be separated from their recombination. This can be achieved using nanocomposite semicondutors. Therefore, the research should be directed to synthesize nanocomposite semiconductors and to study photosplitting of water using nanocomposite semiconductors. The following flow chart gives subsequent reactions for obtaining nanocomposite semiconductors.

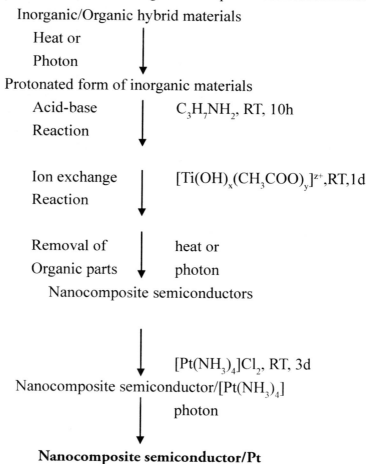

Table 11 summarizes selected semiconductors that find application as photocatalysts. These compounds have been reported in the literature and I would like to propose these semiconductors as starting point to get protonated form by photo decomposition of organic moieties and subsequently they will be explored to get nanocomposite semiconductors.

Table 11: Selected Semiconductors

Compound before decomposition	Photodecomposable moity	Expected Product
$M_4In_{16}S_{33} \cdot (H_2O)_{20} \cdot (C_{10}H_{28}N_4)_{2.5}$ Where M = Cd and Zn	$C_{10}H_{28}N_4$	$M_4In_{16}S_{33} \cdot (H_2O) \cdot (H^+)_{10}$ Where M = Cd and Zn
$[(CH_3CH_2)_2NH_2]_6In_{10}S_{18}$	$[(CH_3CH2)_2NH]$	$In_{10}S_{18}(H^+)_6$
$[C_6H_{16}N]_4In_4S_{10}H_4$	$C_6H_{16}N$	$In_4S_{10}H_4 \cdot (H^+)_4$
$[C_{13}H_{14}N_2]_4 \cdot In_9S_{17}$	$C_{13}H_4N_2$	$In_9S_{17} \cdot (H^+)_7$

DESIGN, SYNTHESIS, AND PHOTOCATALYSIS OF SEMICONDUCTORS WITH OPEN FRAMEWORK STRUCTURE

Exploratory hydrothermal synthesis will be employed to synthesize different open framework compounds under various synthesis conditions.

Hydrothermal synthetic technique employed to synthesize these compounds involves low-temperature heating (110°-250°C) of corresponding metal salt or metal powder with H_3PO_4, H_3BO_3, SiO_2, arsenic acid or oxalic acid in presence of template molecules under autogeneous pressure in Teflon-lined bomb. The template molecules are, in general, diamines. These molecules exhibit extensive hydrogen bonding with framework oxygen and hydroxyl groups. Formation of different compounds and structures is governed mainly by organic templates, reaction temperatures, pH, concentration and solvent.

Semiconductor nanocomposites with layered structure play a vital role in water splitting and detoxification. Thus, the layered oxides synthesized using transition metals are Ti, Nb, W etc. The preparation of these materials involves solid state reaction of corrersponding oxides and/or carbonates. These compounds are then modified by ion-exchange and intercalation methods to get semiconductor nanocomposites. These semiconductor nanocomposites show improved photocatalytic activity.

The proposed research combines open framework materials with semiconducting properties and to employ these open framework semiconductors for photosplitting of water. Open framework and channels (microporous) present in these well-crystalline compounds make their specific surface area large and hence, photosplitting of water is efficient. Synthesis of these open framework semiconductors involves hydrothermal reaction of corresponding metal powder in presence oxidizer under autogeneous pressure.

Table 12: Metal powders with different oxidizer for various semiconductors

Metal powder	Oxidizer	Organic template
Ti Zn W Nb In Sn	H_2O_2 $(NH_4)_2S_2O_3$ $KMnO_4$ $K_2Cr_2O_7$ HNO_3 NH_4ClO_4 $NaBrO_3$ $NaClO_3$	Various simple and branched diamines and hydrazides

Table 12 summarizes possible metal powders and various oxidizers with different organic templates. Under these conditions, metal powder is oxidized by oxidizer into metal oxide. Since diamine or hydrazide is present in the reaction, it behaves as templates for oxides synthesis. Thus, synthesis of diamine molecules templated metal oxides could be achieved. These compounds will be characterized by single or powder XRD, simultaneous TG-DTA, FT-IR, SQUID and DRS. The study of these compounds comprises photosplitting of water using Hg or Xe lamp.

IGNEOUS COMBUSTION SYNTHESIS, CHARACTERIZATION AND VISIBLE LIGHT DRIVEN WATER SPLITTING USING TRANSITION METAL ION INCORPORATED SEMICONDUCTORS

Thrust in finding suitable semiconductor with visible region band gap for photosplitting of water is the world wide interest. Conventional semiconductor is bulk TiO_2, which shows photosplitting of water into H_2 and O_2 under UV irradiation. Later, $K_4Nb_6O_{17}$, having layered structure, was found better photosplitting of water into H_2 and O_2. Thus, research activities have been directed to explore layered semiconductor for photosplitting of water. But, till now, there is no single semiconductor of layered structure found to exhibit photosplitting of water under visible irradiation.

The on-going research in the area of photosplitting of water is doping of foreign elements in semiconductors and preparation of solid solution of semiconductors. Therefore, I would like to propose the solution combustion technique to synthesize various doped semiconductors. The various semiconductors with different dopants explored by us are summarized in table 13.

Table 13: Semiconductors and dopants for different photocatalysts

Proposed system	Compound Semiconductor	Dopant
$ZnO-TiO_2$	$ZnTiO_3$ Zn_2TiO_4 $Zn_2Ti_3O_8$	Cobalt, Manganese, Nickel
$ZnO-V_2O_5$	ZnV_2O_6 Porous $Zn_3(VO_4)_2 \cdot 3H_2O$ $Zn_2V_2O_7$	Cobalt, Manganese, Nickel
$ZnO-SnO_2$	Zn_2SnO_4 $ZnSnO_3$	Cobalt, Manganese, Nickel

MVO_4, M = AL, BI, CR, FE, Y, EU WITH OPEN FRAMEWORK STRUCTURE

After the discovery of microporous $AlPO_4$ (oxides of main group elements) there has been plethora reports in the literature on various transition metal phosphates having open framework structures. PO_4 group is a well-known moiety and it is having tetrahedral geomentry. Therefore, my idea in this direction is to replace PO_4 group by isostructural VO_4 group. Even though $AlVO_4$, $BiVO_4$ and YVO_4 are well known catalysts for hydrocarbon oxidation, photosplitting of water and stable host for rare earth ions respectively, there is not study on microporous metal vanadates. It is also possible to substitute Al in 6-coordination of $AlVO_4$ with Cr^{3+} to get laser material similar to that of ruby.

Hydrothermal technique would be a good technique to get open framework vanadates. Thus, this method includes starting materials of Al, Bi, V, Fe, Y, Eu and Cr metal powders with water and organic diamines as reaction medium and structure directing group respectively. The reaction temperature varies from 100 to 250°C (low temperature hydrothermal technique).

Outcomes of this proposal:
1. Microporous $AlVO_4$ catalyst could be achieved and the high surface area of this material might have higher catalytic activity for hydrocarbon oxidation when compare to that of bulk $AlVO_4$.
2. Microporous $BiVO_4$ might be a better photosplitting of water when this is compared to that of bulk $BiVO_4$.
3. Microporous YVO_4 can be a good host for Eu and hence, this phosphor might be useful in LED solid state lighting to convert long UV into visible emission.
4. Substitution of Al at 6-coordinated in $AlVO_4$ by Cr will be the lowest temperature synthesis of known laser inorganic material.

Relevant References:

Inorganic Chemistry

1. Chemistry, Zumdahl and Zumdahl, Houghton Mifflin Company, USA, 2004, 6th edition, pages 289-335.
2. General Chemistry, N.L. Glinka, Mir publishers, Moscow, 1981, 3rd edition, vol. 2, pages 40-58, 299-308.
3. Textbook of inorganic chemistry, P.L. Soni and Mohan Katyal, Sultan chand & sons, India, 2007, 20th edition, pages 2.578-2.587.

Synthesis of Inorganic Materials

1. S. Ekambaram, Short notes on applied and advanced inorganic materials chemistry, iUniverse, USA, 2006.
2. CNR Rao, Chemical synthesis of solid inorganic materials, *Mater. Sci. Eng.* **B18**, pp.1-21, 1993.
3. CNR Rao, Chemical Approaches to the synthesis inorganic materials, New Delhi, Wiley Eastern Limited, 1994.
4. D. Segal, Chemical synthesis of ceramic materials, *J. Mater. Chem.*, 7, pp.1297-1305, 1997.
5. R. Nagarajan and CNR Rao, Structure and superconducting properties of Ga-substituted $YBa_2Cu_3O_{7-\delta}$ and $YBaSrCu_3O_{7-\delta}$ systems, *J. Mater. Chem.*, **3**, pp. 969-973, 1993.
6. J.S. Yoo, S.H. Kim, Y.T. Yoo, G.Y. Hong, K.P. Kim, J. Rowland and P.H. Holloway, Control of spectral properties of strontium-alkaline earth-silicate-europium phosphors for LED applications, *J. Electrochem. Soc.*, **152**, No. 5, pp.G382-G385, 2005.

7. E.G. Gillan and R.B. Kaner, Rapid, energetic metathesis routes to crystalline metastable phases of zirconium and hafnium dioxide, *J. Mater. Chem.*, **11**, pp. 1951-1956, 2001.
8. T. Sivakumar, J. Gopalakrishnan, Transformation of Dion-Jacobson phase to aurivilies phase: synthesis of (PBBiO$_2$)MNb$_2$O7 (M = La, Bi), *Mater. Res. Bull.*, **40**, pp.39-45, 2005.
9. S. Ekambaram, K.C. Patil and M. Maaza., Synthesis of lamp phosphors:facile combustion approach, *J. Alloys and Compounds*, **393**, pp. 81-92, 2005.
10. K.C. Patil, S.T. Aruna and S. Ekambaram, Combustion synthesis, *Curr. Opin. Solid State mater. Sci.*, **2**, pp. 158-165, 1997.
11. S. Ekambaram and S.C. Sevov, Organically templated mixed valent Ti^{III}/Ti^{IV} phosphate with an octahedral-tetrahedral open framework, *Angew. Chem. Int. Ed.*, **38**, pp. 372-375, 1999.
12. S. Ekambaram, C. Serre, G. Ferey and S.C. Sevov, Hydrothermal synthesis and characterization of an ethylene diamine templated mixed valence titanium phosphate, *Chem. Mater.*, **12**, pp. 444-449, 2000.
13. J.H. Cho, Y.J. Ma, Y.H. Lee, M.P. Chun and B.I. Kim, Piezoelectric ceramic powder synthesis of bismuth sodium titanate by a hydrothermal process, J. Ceram. Proce. Research, **7**, pp. 91-94, 2006.
14. S. Ekambaram et. al., unpublished work.
15. Ulrich Schubert and Nicola Husing, Inorganic materials synthesis, Wiley-VCH, 2004.

16. A.R. Boccaccini and I. Zhitomirsky, Application of electrophoretic and electrolytic deposition techniques in ceramics processing, Current Opinion Solid State and Mater. Sci., **6**, pp. 251-260, 2002.
17. G.H.A. Theresa and P.V. Kamath, Electrochemical synthesis of metal oxides and hydroxides, Chem. Mater. **12**, pp. 1195-1204, 2000.

Some Recent Developments

18. S. Ekambaram, M.Yanagisawa, S. Uchida, Y. Fujishiro and T. Sato, "Synthesis and photocatalytic property of hectorite/(Pt. TiO_2) and $H_4Nb_6O_{17}$/(Pt, TiO_2) nanocomposites", *Mol. Cryst. And Liq. Cryst.*, 2000, **341**, 213-218.
19. CNR. Rao and F.L. Deepak, Absence of ferromagnetism in Mn- and Co- doped ZnO, *J. Mater. Chem.*, **15** pp. 573-578, 2005.
20. K.R. Kittilstved and D.R. Gamelin, Activation of High-Tc ferromagnetism in Mn^{2+} doped ZnO using amines, *J. Am. Chem. Soc.*, **127**, pp. 5292-5293, 2005.
21. R. Mark Ormerod, Solid oxide fuel cells, Chem. Soc. Rev., 2003, **32**, 17.
22. Y. Nishihata, J. Mizuki, T. Akao, H. Tanaka, M. Uenishi, M. Kimura, T. Okamoto and N. Hamada, Self-regerneration of a Pd-perovskite catalyst for automotive emissions control, Nature, 2002, **418**, 164.
23. H. Tanaka and M. Misono, Advances in designing perovskite catalysts, Current Opinion in Solid State and Materials Science, **5**, 2001, 381-387.
24. B.C. Sales, Smaller is cooler, Science, 2002, **295**, 1248.

25. Jun Lin Yuan, Jiao Wang, Ding Bang Xiong, Jing Tai Zhao, Yi Bing Fu, Guo Bin Zhang and Chao Shu Shi, Potential PDP phosphors with strong absorption around 172 nm:Rare earth doped $NaLaP_2O_7$ and $NaGdP_2O_7$, J. Lumin., **126**, pp. 717-722, 2007.
26. Ji-W Mooon, W-S Seo, H. Okabe, T. Okawa, K. Koumoto, Ca-doped $RCoO_3$ (R = Gd, Sm, Nd, Pr) as thermoelectric materials, J. Mater. Chem., 2000, **9**, 2007.
27. T. Justel, H. Nikol and C. Ronda, New developments in the field of luminescent materials for lighting and displays, *Angew. Chem. Int. Ed.*, **37**, pp. 3084-3103, 1998.
28. S. Neeraj, N. Kijima and A.K. Cheetham, Novel red phosphors for solid state lighting the system $Bi_xLn_{1-x}VO4:Eu^{3+}/Sm^{3+}$ (Ln = Y & Gd), *Solid State Commun.*, **131**, pp. 65-69, 2004.
29. S. Neeraj, N. Kijima and A.K. Cheetham, Novel red phosphors for solid-state lighting; the system $NaM(WO_4)_{2-x}(MoO_4)_x:Eu^{3+}$ (M =Gd, Y, Bi), *Chem. Phys. Lett.*, **387**, pp. 2-6, 2004.
30. http://web.ukonline.co.uk/aplr/structure.htm
31. V. Thangadurai and W. Weppner, $Li_6Ala_2Nb_2O_{12}$ (A = Ca, Sr, Ba): A new class of fast lithium ion conductors with garnet like structure J. Amer. Ceram. Soc., **88**, pp. 411-418, 2005.
32. Chang-Hong Kim, II-Eok Kwon, Cheol-Hee Park, Young-Ju Hwang, Hyun-Sook Bae, Byung-Yong Yu, Chong-Hong Pyun and Guang-Yan Hong, Phosphors for plasma display panels, J. Alloys Compounds, **311**, pp 33-39, 2000.

Research Needed in Photosplitting of Water

33. A. Kudo, J. Ceram. Soc. Japan, Development of photocatalyst materials for water splitting aiming at light energy conversion, **109(6)**, 2001, S81-S88.
34. K. Sayama and H. Arakawa, Effect of carbonate salt addition on the photocatalytic decomposition of liquid water over Pt-TiO_2 catalyst, J. Chem. Soc., Faraday Trans., **193**, 1997, 1647-54.
35. S. Ikeda, A. Tanaka, K. Shinohara, M. Hara, J.N. Kondo, K. Maruya and K. Domen, Effect of the particle size for photocatalytic decomposition of water on Ni-loaded $K_4Nb_6O_{17}$, Microporous Materials, **9**, 1997, 253-58.
36. Z. Zou, J. Ye, K. Sayama and H. Arakawa, Direct splitting of water under visible light irradiation with an oxide semiconductor photocatalyst, Nature, **414**, 2001, 625.
37. M. Hartmann and L. Kevan, Transition-metal ions in aluminophosphate and silicoaluminophosphate molecular sieves: Location, Interaction with adsorbates and catalytic properties, Chem. Rev., 1999, **99**, 635.
38. H. Li, J. Kim, T.L. Groy, M. O'Keeffe and O.M. Yaghi, 20Å $Cd_4In_{16}S_{35}^{10-}$ supertetrahedral T4 clusters as building units in decorated cristobalite frameworks, J. Am.Chem. Soc., 2001, **123**, 4867.
39. C. Wang, Y. Li, X. Bu, N. Zheng, O. Zivkovic, C.S. Yang and P. Feng, Three-dimensional superlattices built from $(M_4In_{16}S_{33})^{10-}$ (M = Mn, Co, Zn, Cd) supertetrahedral clusters, J.Am.Chem.Soc., 2001, **123**, 11506
40. C.L. Cahill and J.B. Parise, On the formation of framework indium sulfides, J.Chem.Soc., Dalton Trans., 2000, 1475.

41. A.K. Cheetham, G. Ferey and T. Loiseau, Open-framework Inorganic Materials, Angew. Chem. Int. Ed., 1999, **38**,3269.
42. J.M. Thomas and W.J. Thomas, Principles and practice of heterogeneous catalysis, VCH, NY, USA.

Appendix-1:
Michael Faraday (1791-1867)

Faraday was born on September 22, 1791 in England. He was in a poor family those days and he did not complete education in his childhood. He found a job as an apprentice in a bookseller and book binder shop. In this job he utilized the opportunity of reading scientific articles, especially on electricity and magnetism.

In 1812 he attended series of lectures given by the british chemist, Sir Humphry Dave. These lectures changed the life of Faraday. He made notes on those lectures and as binded book Faraday sent it to Sir Humphry Dave to seeking for a job in his laboratory. Finally, Faraday was appointed as assistant in Davy's laboratory. In fact, Faraday's scientific work started in Royal Institution in London where Davy was a principal investigator in chemistry.

Those days scientists were investigating electricity and magnetism. Therefore, Faraday paid his attention towards electricity and magnetism and in fact, he observed a breakthrough in his experiments. The breakthrough was generation of electricity by motion of magnetic material. This result made Faraday to be well-known in the world and he was the first scientist in the world to generate electricity by motion of magnetic materials. Also, Faraday invented electrochemistry where in his contribution was laws of electrolysis. Even today, his laws are found in textbooks of electrochemistry.

Thomas Alva Edison (1847-1931):

Thomas Alva Edison was born to sam and nancy on Feb. 11, 1847 in ohio, USA and he was the youngest child of seven children. He did not have proper school education but his mother took care of Edison for his early education. He started to work at the age of 12 to sell newspaper and candy and meantime, he set up a small laboratory in the baggage car.

Even when Edison was a young child Edison wanted to carry out experiments to understand the nature. There were two of them very notable. The first one was the hatching eggs by sitting on them since Edison saw chicken to hatch eggs by sitting on them. This experiment was failed. The second experiment was interesting one. Why not human could not fly like birds? This was a question in Edison's mind those days. Therefore, he carefully observed birds and Edison found that birds ate worm as a main food. Edison thought that if human being ate worm, then, he could flew. In order to experiment this idea, he gave smashed worm liquid to his friend to drink and to see whether or not his friend flew. But, this experiment was also failed.

To his credit, there were numerous patents for him, which is the world record even now. He failed 1000 times during his achievement to make a cheap electric bulb. Many scientists investigated to make cheap electric lamp. But, most of them left the work and told the world that it could not be possible. But, Edison continued to work on this project and finally, he achieved to make cheap electric lamp. Edison had other inventions such as phonograph and motion picture to name a couple.

CPSIA information can be obtained at www.ICGtesting.com
Printed in the USA
BVOW072008100713

325648BV00003B/221/P